Anxieties and Management Responses in International Business

THE ACADEMY OF INTERNATIONAL BUSINESS

Published in association with the UK & Ireland Chapter of the Academy of International Business

Titles already published in the series:

International Business and Europe in Transition (Volume 1)
Edited by Fred Burton, Mo Yamin and Stephen Young

Internationalisation Strategies (Volume 2)
Edited by George Chryssochoidis, Carla Millar and Jeremy Clegg

The Strategy and Organization of International Business (Volume 3)
Edited by Peter J. Buckley, Fred Burton and Hafiz Mirza

Internationalization: Process, Context and Markets (Volume 4)
Edited by Graham Hooley, Ray Loveridge and David Wilson

International Business Organization (Volume 5)
Edited by Fred Burton, Malcolm Chapman and Adam Cross

International Business: Emerging Issues and Emerging Markets (Volume 6)
Edited by Carla C. J. M. Millar, Robert M. Grant and Chong Ju Choi

International Business: European Dimensions (Volume 7)
Edited by Michael D. Hughes and James H. Taggart

Multinationals in a New Era: International Strategy and Management (Volume 8)
Edited by James H. Taggart, Maureen Berry and Michael McDermott

International Business (Volume 9)
Edited by Frank McDonald, Heinz Tüselmann and Colin Wheeler

Internationalization: Firm Strategies and Management (Volume 10)
Edited by Colin Wheeler, Frank McDonald and Irene Greaves

The Process of Internationalization (Volume 11)
Edited by Frank McDonald, Michael Mayer and Trevor Buck

International Business in an Enlarging Europe (Volume 12)
Edited by Trevor Morrow, Sharon Loane, Jim Bell and Colin Wheeler

Managerial Issues in International Business (Volume 13)
Edited by Felicia M. Fai and Eleanor J. Morgan

Anxieties and Management Responses in International Business (Volume 14)
Edited by Rudolf R. Sinkovics and Mo Yamin

Anxieties and Management Responses in International Business

Edited by

Rudolf R. Sinkovics

and

Mo Yamin

First published 2007 by
PALGRAVE MACMILLAN
Houndmills, Basingstoke, Hampshire RG21 6XS and
175 Fifth Avenue, New York, N.Y. 10010
Companies and representatives throughout the world

PALGRAVE MACMILLAN is the global academic imprint of the Palgrave
Macmillan division of St. Martin's Press, LLC and of Palgrave Macmillan Ltd.
Macmillan® is a registered trademark in the United States, United Kingdom
and other countries. Palgrave is a registered trademark in the European
Union and other countries.

ISBN-13: 978–0–230–51556–7
ISBN-10: 0–230–51556–8

This book is printed on paper suitable for recycling and made from fully
managed and sustained forest sources.

A catalogue record for this book is available from the British Library.

Library of Congress Cataloging-in-Publication Data
Anxieties and management responses in international business / edited
    by Rudolf R. Sinkovics and Mo Yamin.
      p.   cm.
    Includes bibliographical references and index.
    ISBN-13: 978–0–230–51556–7 (cloth)
    ISBN-10: 0–230–51556–8 (cloth)
    1. International business enterprises—Management.   2. International
    business enterprises—Planning.   3. Strategic planning.   I. Sinkovics,
    Rudolf R., 1966–   II. Mo, Yamin.
    HD62.4A49 2007
    658'.049—dc22   2006052475

10   9   8   7   6   5   4   3   2
16   15   14   13   12   11   10   09   08   07

Printed and bound in Great Britain by
Antony Rowe Ltd, Chippenham and Eastbourne

Contents

List of Tables

List of Figures

Foreword

The 33rd Annual Conference of the Academy of International Business, United Kingdom Chapter, was hosted by Manchester Business School, University of Manchester, on Friday 7th and Saturday 8th April 2006. This volume is the fourteenth to be published in the Palgrave Macmillan series of selected papers from the annual conference – now under the series editorship of Michael Mayer, of the University of Bath. The UK (now UK & Ireland) Chapter has the distinction of being the only chapter of the Academy of International Business to publish a volume of papers from its annual conference.

The organizers of the 2006 conference were Pervez N. Ghauri (Conference Chair), Rudolf R. Sinkovics and Mo Yamin, with managing support by Robert-Jan F. G. Bulter, and Conference Secretary Gill Geraghty. The conference was the largest to date in the history of the Chapter. More than 125 papers were presented. There were 212 participants, with 112 from UK universities (staff and research students), i.e. just over half of those attending. The other participants came from 26 countries outside the UK. This gave the meeting a truly international feel. A key part of the annual conference is the Doctoral Colloquium, for those students pursuing research degrees, organized by the Doctoral Convenor, Marian Jones of the University of Glasgow. A record 44 students registered to present their work: 19 students in the stream for those early on in their research, and 25 in the more advanced ('Competitive') stream.

The Opening Keynote Session of the Conference was chaired by Peter Buckley, and comprised presentations by two invited speakers: Professor S. Tamer Cavusgil, of Michigan State University, USA, and Professor Mats Forsgren of Uppsala University Sweden. Professor Cavusgil spoke on 'International Business in the Age of Anxiety: Company Risks Arising from Poor Partnering', and Professor Forsgren took as his subject 'Are Multinationals Superior or Just Powerful? A Critical Review of the Evolutionary Theory of MNCs'. The second plenary session, 'International Business in the Age of Anxiety: Four Perspectives', was an innovative bringing together of presentations from academia, business, government and international organizations. The presenters were John Dunning, Emeritus Professor of International Business at the University of Reading, UK, and State of New Jersey Emeritus Professor of International Business at Rutgers University, Peter A. Harwood, Allied Domecq West, John Doddrell, Director, Strategy & Human Resources, UK Trade & Investment, and Hafiz Mirza, Chief, Development Issues Section, Division on Investment, Technology and Enterprise Development, United Nations.

The overarching theme 'International Business in the Age of Anxiety' was found to be a strong and rich one throughout the Conference. In this volume, discussion of a number of sources of (and responses to) anxiety runs through each of the sections. As it also would happen, the academic study of international business is itself thought by some to be experiencing its own bout of anxiety. There are questions over the place of international business-focused research in a world in which the disciplines that make up business and management are themselves increasingly international in orientation. What, then, is the role of research in international business? This volume does something to answer this. It shows the value of an approach that relies not on one discipline, but on many; of one that addresses topics that do not fall neatly into conventional categories of enquiry; and of one that does not take a single country's experience (or that of a limited set) as representative of the entire world. The research within these pages shows that deep academic enquiry thrives on inspiration from outside the narrow disciplines, and is at its best when it strives to be relevant to the world outside the academic community.

In 2006, the United Kingdom & Ireland Chapter was formed as a 'multi-country chapter' according to the Constitution of the world-wide Academy. This brought together the AIB membership of the United Kingdom Chapter – established in 1973 at the University of Manchester Institute of Science and Technology by founding chairman Dr Michael Z. Brooke – and the AIB membership in the Republic of Ireland. The Chapter, with a record membership of 314 as of June 2006, shows no sign of succumbing to anxiety, and is in a stronger position than ever before to promote the research and study of international business in the UK and Ireland. This stimulating collection of papers is powerful evidence of the Chapter's exciting international business agenda.

JEREMY CLEGG
Chair, Academy of International Business
United Kingdom & Ireland Chapter

Notes on the Contributors

Kirsimarja Blomqvist is a Professor of Knowledge Management at Lappeenranta University of Technology, Finland. Her research interests include trust and social capital in knowledge creation and innovation; R&D and innovation management; and alliances and networks. She has published several books, contributed to others, and her work has been published in peer-reviewed journals.

Keith D. Brouthers is Professor of Business Strategy at King's College London, UK. Before becoming an academic, he spent twelve years in the business world, first as a certified public accountant (CPA) and certified management accountant (CMA), then as Chief Financial Officer in several international companies. He has published widely in the area of international entry mode choice and strategic decision-making. His work also examines issues of joint-venture management.

Mark Casson is Professor of Economics and Director of the Centre for Institutional Performance at the University of Reading, UK. His research interests include entrepreneurship and business history as well as international business studies.

Francesco Ciabuschi is Lecturer in International Business at the Department of Business Studies, Uppsala University, Sweden. He has published on strategy, IT and international business issues, and is an active member of several academic associations and work groups. He received his PhD from the Department of Business Studies, Uppsala University.

Pavlos Dimitratos is Lecturer in International Business at the Athens University of Economics and Business, Greece, and a visiting Senior Research Fellow at the University of Glasgow, UK. His research interests include international management and strategy, and entrepreneurship. He has published in several international business, entrepreneurship and management journals. He received his PhD from the Manchester Business School, UK.

Fred van Eenennaam is Professor of Strategy and Dynamics of Strategy at Nyenrode Business Universiteit, The Netherlands. He has a family business background, and business and consulting experience in various industries. His research interests are international business, dynamics of strategy, corporate governance and strategic alignment. He holds a PhD from the Vrije Universiteit Amsterdam and his dissertation topic was 'European Market-based Strategies of Consumer Goods Companies'.

Mats Forsgren is Professor in International Business at Uppsala University, Sweden. He has also been a Professor at Copenhagen Business School, a visiting Professor at Stockholm School of Economics, and a Research Fellow at the Swedish School of Economics in Helsinki, Finland. His publications include over fifty articles and books concerning foreign direct investment theory, the internationalization process, managing the multinational firm, and network theory. He is an Honorary Doctor at the Swedish School of Economics, Helsinki.

Fabienne Fortanier is a PhD candidate at the University of Amsterdam Business School (ABS). Her research focuses on the economic and social development impact of inward foreign direct investment, with a particular emphasis on the role of multinational firms' strategies.

Pervez N. Ghauri completed his PhD at Uppsala University, Sweden, where he also taught for some years. At present he is Professor of International Business at Manchester Business School in the UK. Professor Ghauri specializes in entry strategies and international business negotiations. He has published more than twenty books and numerous articles on international marketing and international business issues. He is also editor-in-chief of *International Business Review*, the official journal of the European International Business Academy (EIBA).

Damian P. Grimshaw is Professor in Employment Studies at Manchester Business School, UK, and Director of the European Work and Employment Research Centre (EWERC). His research covers several areas of employment policy and practice, involving both case-study research in small and large organisations and cross-national comparisons of employment systems. He has a BSc in Maths and Management Science from UMIST, an MA in Economics from the University of Massachusetts at Amherst, and a PhD from UMIST.

Amjad Hadjikhani is Professor of Marketing in the Department of Business Studies at Uppsala University, Sweden. He has been actively engaged in research work, and has published several books and numerous articles on international marketing, in particular on consumer and industrial marketing.

Jussi Hätönen is a researcher in the Department of International Business at the Turku School of Economics, Finland. His research focuses on international outsourcing, in particular in the context of the software and ICT industry. He is currently pursuing a PhD in this area.

Matthias Hossinger gained an MSc in International Business and Management from Manchester Business School, UK. His MSc dissertation research dealt with website adaptation and standardization strategies of German multinationals. He now works in an IT function with Hilti (Schweiz) AG in Switzerland.

Claudia M. L. Janssen gained her MSc in Business Economics from the Vrije Universiteit Amsterdam. She works as lecturer at the Strategy Center of Nyenrode Business Universiteit and as consultant in the tax decision management team at Deloitte Amsterdam. Her PhD research is in the field of strategy and decision-making.

Nada Korac Kakabadse is Professor of Management and Business Research at the Northampton Business School, UK. She is co-editor of the *Journal of Management Development* and editor of the *Corporate Governance* journal. She has published widely in the areas of strategic leadership, IS/IT policy, governance, boardroom effectiveness, diversity management and ethics, including six books and over eighty scholarly and reviewed articles, and acted as consultant to numerous organizations.

Ans Kolk is Professor of Sustainable Management and Research Director of the University of Amsterdam Business School (ABS). Her areas of research and publications are in corporate social responsibility and environmental management, in particular in relation to multinational corporations' strategies, international policy and development issues.

Olli Kuivalainen is Professor of International Marketing at the Lappeenranta University of Technology (LUT), Finland. His research interests are in the areas of internationalization and growth strategies of knowledge-intensive firms, international marketing, and international entrepreneurship. He received his doctoral degree in Economics and Business Administration from the LUT, Finland.

Amanda Jane Langley is a lecturer at Northampton Business School, UK. Her research interests include strategic action coevolution, strategic change and the development of qualitative data analysis approaches for understanding strategic dynamics within industries.

Joong-Woo Lee is Professor of International Business in the School of Management, Inje University, Korea. He completed his PhD at Uppsala University, Sweden and has conducted research in Korea. He is involved in IB research covering firms in South East Asia, and has published several books and a number of articles on industrial marketing.

Jani Lindqvist is a researcher in international marketing at Lappeenranta University of Technology, Finland. His primary research interests comprise internationalization strategies and the partnering of small, service-orientated ICT firms.

Rod B. McNaughton holds the Eyton Chair in Entrepreneurship at the University of Waterloo, Canada. His research focuses on the growth of knowledge-based ventures, especially the development of a market orientation, early internationalization, financing, and their roles within industrial clusters. He received his PhD in Marketing from Lancaster University, UK, and his PhD in Economic Geography from the University of Western Ontario, Canada.

Marcela Miozzo is Professor of Economics and Management Innovation at Manchester Business School, UK. Her main interests are in internationalization and innovation in services, innovation in construction, and innovation and firm capabilities in less developed countries. She completed a degree in Economics at the University of Buenos Aires, Argentina, an MA in Technology Policy and Innovation Management at MERIT, Maastricht, The Netherlands, and a PhD in Economics at the University of Massachusetts, Amherst, USA.

Elfriede Penz is Assistant Professor at the Institute of International Marketing & Management at Wirtschaftsuniversität Wien (WU-Wien), Austria. She gained a Ph.D. in Psychology from the University of Vienna and a European PhD on Social Representations and Communications from the Universities 'La Sapienza' in Rome, 'ISCTE' in Lisbon and in Helsinki. Elfriede Penz also holds a Master of Advanced Studies, degree as Cultural Manager of the University of Music and Performing Arts, Vienna. Research interests include consumer behaviour as well as methodological issues in international marketing and management.

Mika Ruokonen is a researcher in the Technology Business Research Center at Lappeenranta University of Technology, Finland. His research interests are rapid internationalization of small technology-based firms and their business networks in the internationalization process. He has published on these issues in the *International Journal of Entrepreneurship* and *Innovation Management*.

Sami Saarenketo is Professor of International Marketing at Lappeenranta University of Technology, Finland. His primary areas of research interest are internationalization, international marketing and entrepreneurship in knowledge-intensive firms. He has published on these issues in the *Canadian Journal of Administrative Sciences*, and the *International Journal of Production Economics*, among others. He received his doctoral degree from the Lappeenranta University of Technology.

Helen Salavou is a visiting lecturer at the Athens University of Economics and Business, Athens, Greece. Her research interests involve the areas of product innovation, entrepreneurship and small business research. She has published in the *European Journal of Marketing, European Journal of Innovation Management*, and *Creativity and Innovation Management*. She received her PhD from the Athens University of Economics and Business.

Per Servais is Senior Lecturer in International Marketing Strategy in the Department of Marketing and Management, University of Southern Denmark, His research topics are industrial marketing, international entrepreneurship and SME internationalization. He received his PhD in International Business from Odense University, Denmark.

Rudolf R. Sinkovics is Senior Lecturer in International Business at Manchester Business School, UK. His research centres on inter-organizational governance, the role of ICT in firms' internationalization, and research methods in international business. He received his PhD from Vienna University of Economics and Business Administration (WU-Wien), Austria.

Sanna Sundqvist is Professor of International Marketing in the Department of Business Administration, Lappeenranta University of Technology (LUT), Finland. Her research interests are on export performance, diffusion of innovations, and market orientation, and she has published widely in these areas. She received her DSc in Technology from LUT.

Stephen Swailes is in the Centre for Organisational Learning at the University of Hull, UK. His interests cover organizational behaviour in technology industries and the work behaviour of professional employees.

Irini Voudouris is Lecturer in Technology and Management at the Athens University of Economics and Business, Athens, Greece. Her research interests involve the areas of strategic management, entrepreneurship, technology and management, and flexibility of organizations. She has published in the *Journal of Human Resource Management* and *European Management Journal*, among others. She received her PhD from the Université Paris X-Nanterre, France.

Mo Yamin is Reader in International Business at Manchester Business School, UK. His research focuses on the role of subsidiaries within multinational companies, linkages of firms with local companies and the impact of multinational companies on economic development. He received his PhD in Economics from the University of Manchester, UK.

Nan Sheng Zhang is a graduate of the MLitt programme, School of Management, St Andrews University, Scotland.

Introduction

Mo Yamin and Rudolf R. Sinkovics

The theme of this volume is 'anxieties' in international business and their managerial ramifications. The dictionary definition of anxiety highlights 'doubt concerning the reality and nature of the threat', and 'self-doubt about one's capacity to cope with it' (Merriam-Webster OnLine, 2006). The environment of international business has always been simultaneously complex, competitive and dynamic. Much discussion in the international business (IB) literature is routinely prefaced by a recognition of the 'challenging' nature of the managerial tasks. In the current context, the image of anxiety is mainly associated with the after-effects of the tragic events of 9/11 (September 11, 2001, World Trade Center, New York) and the heightened concerns with security as a major feature of the macro environment of international business (Kotabe, 2005; Prasad and Ghauri, 2004). However international business has arguably always been beset with anxieties stemming from varied and changing environmental conditions, and the complexities and risks associated with operating across distances and across cultural, institutional, political and economic domains. In fact, it is not an exaggeration to suggest that international business is 'normally' affected by anxiety to a greater degree than 'domestic' or national business, although the boundaries between these categories are increasingly porous.

A key actor in the international business environment is the multinational enterprise (MNE), and one can make the case that the organization and politics of the MNE provide a potential pool of anxiety. Thus 'autonomous 'subsidiaries may well experience uncertainty regarding the exact limits to their authority, while headquarters are continually anxious about whether and how much control they in fact exert on various subunits. Expatriate managers suffer anxiety that being away from the centre may damage their network or social capital within the organization's centres of power, thus undermining their future career prospects (Goodall and Roberts, 2003)

The 'drivers' of competitive advantage for all firms, whether MNEs or small and medium enterprises (SMEs), are rarely constant or transparent. The literature suggests that 'causal ambiguity' is a source of sustainable competitive advantage (Dierickx *et al.*, 1989), but this assumes that the causes of a firm's competitiveness are only ambiguous to its rivals; the firm itself is assumed to have a clear and unambiguous understanding. This may be an over-optimistic assumption, as competitive 'recipes' are rarely transferable from one context to another (Jensen and Szulanski 2004). Moreover, the key to competitiveness lies somewhere between resources, networks, relational capital or 'coopetition' (Dunning and Narula, 2004; Luo, 2005) all in a context in which global and local/regional processes interact in complex and non-transparent ways. Thus the process of developing appropriate skills and 'getting things right' is increasingly beset with uncertainties. The image of 'driving' to competitiveness, implying a clearly-aimed and unequivocal process, is grossly misleading.

Anxieties are also manifest from the perspectives of countries and localities impacted by multinational corporations' (MNCs) activities and investment. As MNEs follow ever more 'flexible' strategies (Buckley and Ghauri, 2004) recipient countries, particularly less developed and emerging ones, need to deal with ever more footloose operations and are challenged to increase the stickiness of MNE operations in their countries. Arguably, many governments are investing inordinate amounts of effort and resources in 'marketing' their countries to increasingly choosy MNEs. The opportunity cost of this is measured in terms of reduced funding and policy efforts that can be devoted to discharging the government's other responsibilities (Yamin and Ghauri, 2004). The IB literature has thus far paid inadequate attention to this issue (Wells, 2003).

In this book we have collected together fourteen papers, each becoming an individual chapter, and the book is divided into four parts. All the contributions highlight the complexities of the international business environment or the managerial implications of such complexity.

Part I Networks and subsidiaries

This section contains three chapters. In Chapter 1, Mark Casson provides an illuminating overview of variegated and interconnected networks in the physical and social domains. Networks are a highly pervasive feature of the economy and often ostensibly 'individualistic' activities such as entrepreneurship are in fact highly network-dependent. Casson sets out the main dimensions of network structure, and explores the factors that determine the structure of any given type of network. The chapter considers social networks, commercial networks involving trade and investment, and local business networks, such as those found in industrial districts. It is argued that the structure of a network can usually be understood as the efficient solution to a co-ordination problem. As a result, explanations can be developed of why varying network structures emerge in different situations.

The second chapter, by Francesco Ciabuschi and Mats Forsgren, is set in an intra-MNE network context and focuses on subsidiary entrepreneurship. It is a detailed consideration of a specific innovation (ThermoSafe). The key contribution of this chapter is that it reveals the ambiguous effects of autonomy. Subunit autonomy is often proposed as an important driver of subunit initiative-taking. However, the paper shows that when autonomy is the 'default' outcome of the centre's ignorance or lack of interest in the subsidiary, rather than being a deliberate and negotiated arrangement between the centre and the subsidiary, any positive effect on entrepreneurship may be distorted. Thus subunits managers might, figuratively, continually 'look over their shoulders' in case of a sudden and arbitrary change in the parent's stance *vis-à-vis* the subsidiary. As the authors point out, the freedom to act based on a high degree of perceived autonomy is circumscribed by the high uncertainty linked to unclear rules of the game. Autonomy as a dimension of entrepreneurial orientation is, therefore, a double-edged sword, depending on what the perceived autonomy is, in fact, based.

The third and final chapter in this part, by Jani Lindqvist, Kirsimarja Blomqvist and Sami Saarenketo, considers the role of sales subsidiaries in MNC innovativeness. Although there is much emphasis in the literature on the beneficial affects of customer/market orientation for innovation in organizations, the potential role of sales subsidiaries on MNC innovations is largely ignored. Sales subsidiaries are rarely studied from this perspective. The current study makes a valuable contribution by constructing a conceptual framework and providing interesting insights on the role of sales subsidiaries from five focus group studies (at the HQ and in four sales subsidiaries) on a major MNE. The most interesting findings of Chapter 3 is the perception among sales subsidiary respondents that while their interactions with customers often generate valuable knowledge and are potential sources of innovations, the hierarchical communication structure effectively undervalues this. This suggests that organizational distance between the sales subsidiary and their HQ causes much frustration; communication with the centre is rarely experienced as either mutual or satisfactory.

Part II   International businesses and local market interactions and impacts

This part comprises three chapters. Chapter 4, by Fred van Eenennaam, Claudia M. L. Janssen and Keith D. Brouthers, revisits the traditional question in international marketing – standardization or adaptation – by applying a theoretical framework informed by the construct of institutional distance. Institutional theory arguably provides a richer and more nuanced characterization of local market diversities than do traditional marketing approaches.

Furthermore, institutional theory mandates a discriminating approach reflecting the differential impact of institutional dimensions (regulative, normative and cognitive). The authors show that even though marketing strategy is increasing global in tone – at least, to judge by the pronouncement of chief executive officers (CEOs), institutional differences do have an impact on the ability of the business units to pursue standardization in process or product strategies. The study indicates that the subsidiaries' accommodation to the institutional environment in which it is placed has a significant effect as to the degree to which the centre's strategies can in fact be standardized.

Chapter 5, by Marcela Miozzo and Damian Grimshaw, examines the effects for client firms in (middle-income), less developed countries of outsourcing business functions to new kinds of global service suppliers. Drawing on case studies of information technology (IT) outsourcing in Argentina and Brazil, the chapter shows that the consequences of forward linkages with global services suppliers reflect particular economic and institutional conditions prevalent in these countries. The ability of clients to benefit from linkages is contingent on their absorptive capacity, in particular their expertise in designing and operating IT outsourcing contracts. A detrimental effect on linkages is that IT service suppliers are able to move not only their own operations between countries but also the execution of contracts with clients. These practices relocate clients' outsourcing from subsidiaries of suppliers initially located within a domestic economy to subsidiaries located outside it, facilitating the consolidation and regionalization of business segments of (multinational) clients, making client firms themselves more 'footloose'.

The final chapter in this section, Chapter 6, is by Fabienne Fortanier and Ans Kolk. They consider firms' self-reported impact of their economic activities on society, employment and employee issues. As the authors aptly observe, the business of business is no longer just business. Businesses are also faced with increased public concerns regarding the economic, social and environmental consequences of their activities. As central actors in the economies of both developed and developing countries, the global operations of multinational enterprises are watched and scrutinized increasingly by governments and policy-makers; non-governmental organizations (NGOs), such as trade unions or consumer organizations; and the public at large. Arguably, public perception of MNEs as 'good' global citizens is a great asset, while mounting public concern and dissatisfaction with corporate behaviour and impact may undermine their long-term viability. In this context, it is interesting and useful to examine what perception the MNEs themselves have of their impact. While the extant literature has examined corporate reporting on their environmental (and to a lesser extent, ethical) practices, relatively little attention has been paid to how firms perceive their socio-economic impact. The authors make a strong case for an analysis of firms' self-perception and cognition as being helpful for further research, and for managers and policy-makers interested in assessing and guiding MNE behaviour.

Part III Political and strategic international business challenges

This section contains four chapters. In Chapter 7, Joong-Woo Lee, Pervez N. Ghauri and Amjad Hadjikhani present an analysis of how market entry and development is a process in which business and political factors interact to shape opportunities and outcomes. Business network studies have explored extensively relationships between firms and other business actors, but rarely touched upon the socio-political relationships. Lee *et al.* address this shortcoming via a business network approach, and examine how MNCs manage their relationship with socio-political organizations, to strengthen their position in the international market. The chapter develops a theoretical framework stressing the three interrelated concepts of legitimacy, commitment and trust, which describe the firms' behaviour. The framework is applied in a case study of Daewoo Motor Company, a Korean MNC in Poland and the European Union (EU) market.

Amanda Jane Langley, Nada Korac Kakabadse and Stephen Swailes focus in Chapter 8 on the pharmaceutical industry and consider the internationalization strategies and outcomes for small and medium-sized firms. The authors apply an evolutionary theory to explore the question, 'How did strategic actions regarding the internationalization of incumbent firms evolve as the pharmaceutical industry transformed itself into a global oligopoly?' The pharmaceutical industry was chosen as the focus of study because of changes with regard to both its structure and the pressure that firms face with increased health care reforms. The authors show that, despite all the firms in the sample being medium-sized, with three of them sharing very similar characteristics, they evolved with unique patterns of internationalization outcomes. The patterns of strategic action suggest that a 'mimetic isomorphism' process may be at work. Firms have become increasingly institutionalized within the environment, which has led to a process where they copy the strategic actions of other firms. However the outcomes of strategies in terms of growth and survival were not strongly related to such institutionalization.

Chapter 9, by Elfriede Penz, is an examination of counterfeiting in international business. The chapter aims to contribute to the development of anti-counterfeiting strategies by studying actions that have already been applied in practice to combat counterfeiting by multinational companies. The author argues that the literature has not yet suggested a consistent strategy, and that there is a lack of guidance in the fight against counterfeiting. The study examines responses to counterfeiting from a business network perspective looking at business–consumer, business–business and business–government interactions. The qualitative text-based study includes interviews with managers from multinational companies along with other company documents and revealed insights into their understanding of counterfeiting and their attempts at problem-solving. Results point to industry- and market-specific approaches rather than universal moves in the fight against counterfeiting.

The final chapter in this section, by Jussi Hätönen and Mika Ruokonen, considers the issue of outsourcing. As national barriers diminish, communications links improve and focused supplier bases evolve, outsourcing has become an ever more central part of internationalization. While some scholars have suggested that cross-border outsourcing is the new topic of future IB research, the authors contend that the research community has in fact paid only limited attention to this phenomenon. Accordingly, their chapter illustrates that several aspects of the phenomenon have been overlooked by IB scholars.

Part IV SME internationalization, entrepreneurship and the Internet

The final section consists of four chapters. Pavlos Dimitratos, Irini Voudouris and Helen Salavou examine in Chapter 11 the internationalization of SMEs from relatively peripheral economies, and ask two research questions: how an entrepreneurial firm in a small country on the periphery of the European Union (EU), such as Greece, evaluates and acts on opportunities that lead to growth in international and domestic marketplaces; and how this entrepreneurial firm mobilizes resources in order to expand under harsh environmental conditions and 'anxiety' in its industrial sector. The chapter reports on a longitudinal single case study which provides in-depth evidence on the theme of identification and exploitation of opportunities, and associated organizational growth, which forms an emerging issue of interest in the field of international entrepreneurship.

Chapter 12 is by Matthias Hossinger, Rudolf R. Sinkovics and Mo Yamin. It considers cultural adaptation on the Internet. While some have argued that web-based marketing and internationalization is inherently standardized, there are also compelling reasons for considering that web presences will be culturally adapted. This chapter explores the cultural dimension of an online web presence for German MNEs, and develops hypotheses positing how cultural adaptations may be related to cultural differences between German, US, UK and Latin American host markets. The paper builds on Hofstede's and Hall's cultural framework (Hofstede, 1984, 1991; Hall, 1976). Findings suggest that cultural value depiction is not very strong in the relevant markets, thus a certain degree of 'cultural alientation' occurs. It is suggested that a transnational model of technological knowledge transfer and website management may yield better cultural congruency.

Chapter 13 is by Olli Kuivalainen, Sanna Sundqvist and Per Servais. It examines the geographical dimension in Born Global firms. Despite the increase in studies on Born Globals, there has been little research into the effect of operating in many countries. In many studies, it is only noted that Born Globals derive their turnover from multiple countries, but this 'country effect' has been excluded from the empirical work. The present study explores the global diversity among experienced Born Global firms and attempts to identify drivers and consequences of global diversification strategy. The empirical part of the chapter is based on a survey among Finnish exporting firms, and the results show that both internal and external drivers, such as proactiveness and competitive turbulence, have an effect on global diversity. Rather surprisingly, no significant differences were detected in performance indicators.

The final chapter is by Nan Sheng Zhang and Rod B. McNaughton, and investigates the outcomes of unsolicited international enquiries received by SMEs. The literature on SME internationalization identifies unsolicited international enquiries as an important stimulus to exporting; over 40 per cent of firms initiate their international business based on unsolicited enquiries. Despite this, little research has been carried out to evaluate the quality of unsolicited enquiries in terms of their outcomes – that is, of completed sale results. This chapter reports the experiences of sixty-eight SMEs based in Scotland. The results suggest that, while unsolicited enquiries may be important in increasing awareness of export opportunities, these self-identified customers are not necessarily the best starting point for export activities. In the majority of cases, filling an unsolicited order resulted in some degree of dissatisfaction with the transaction outcome.

References

Buckley, Peter J. and Pervez N. Ghauri (2004) 'Globalisation, Economic Geography and the Strategy of Multinational Enterprises', *Journal of International Business Studies*, 35(2), 81–98.

Dierickx, Ingemar, Karel Cool and Jay B. Barney (1989) 'Asset Stock Accumulation and the Sustainability of Competitive Advantage', *Management Science*, 35(12), 1504–13.

Dunning, John H. and Rajneesh Narula (2004) *Multinationals and Industrial Competitiveness: A New Agenda* (New Horizons in International Business) (Cheltenham: Edward Elgar).

Goodall, Keith and John Roberts (2003) 'Repairing Managerial Knowledge-Ability over Distance', *Organization Studies*, 24(7), 1153–75.

Hall, Edward Twitchell (1976) *Beyond Culture* (New York: Anchor Press).

Hofstede, Geert (1984) *Culture's Consequences: International Differences in Work-Related Values* (Cross-Cultural Research and Methodology Series) (Newbury Park, CA: Sage Publications).

Hofstede, Geert (1991) *Cultures and Organizations: Software of the Mind* (Maidenhead: McGraw-Hill).

Jensen, Robert and Gabriel Szulanski (2004) 'Stickiness and the Adaptation of Organizational Practices in Cross-Border Knowledge Transfers', *Journal of International Business Studies*, 35(6), 508–23.

Kotabe, Masaaki (2005) 'Global Security Risks and International Competitiveness', *Journal of International Management*, 11(4), 453–5.

Luo, Yadong (2005) 'Toward Coopetition within a Multinational Enterprise: A Perspective from Foreign Subsidiaries', *Journal of World Business*, 40(1), 71–90.

Merriam-Webster OnLine (2006) *Anxiety*. Available at http://www.m-w.com/ (accessed 8 August 2006).

Prasad, Benjamin S. and Pervez N. Ghauri (eds) (2004) Global Firms and Emerging Markets in the Age of Anxiety (New York: Praeger).

Wells, Louis T., Jr. (2003) 'Multinationals and the Developing Countries', in Thomas L. Brewer, Stephen Young, and Stephen E. Guisinger (eds), *The New Economic Analysis of Multinationals: An Agenda for Management, Policy, and Research* (Cheltenham: Edward Elgar), 106–21.

Yamin, Mo and Pervez N. Ghauri (2004) 'Rethinking MNE-Emerging Market Relationships: Some Insights from East Asia', in Benjamin S. Prasad and Pervez N. Ghauri (eds), *Global Firms and Emerging Markets in the Age of Anxiety* (Westport, Conn.: Praeger), 251–66.

Part I
Networks and Subsidiaries

1

Networks: A New Paradigm in International Business History?

Mark Casson

Introduction

The concept of a network is now widely used in international business history (IBH). This chapter sets out the main dimensions of network structure, and explores the factors which determine the structure of any given type of network. It considers social networks, commercial networks involving trade and investment, and local business networks, such as those found in industrial districts. It is argued that that the structure of a network can usually be understood as the efficient solution to a co-ordination problem. As a result, explanations can be developed of why different network structures emerge in different situations.

Networks as an inter-disciplinary subject

The concept of a network has become extremely popular in international business history (IBH) (Jones, 2000; Jones and Amatori, 2003). To some writers, it is a unifying paradigm, around which an integrated social science can be built, but to others it is just a passing fad. This chapter argues that the concept of a network is indeed an emerging paradigm, but that the new paradigm will be successful only if researchers can agree on appropriate definitions of terms.

Networks are a powerful way of understanding the historical evolution of institutions. Institutions are often classified using a threefold distinction between firms, markets and the state (North, 1981). Networks are then introduced as a fourth type of institution, with the claim that, until recently, their significance was overlooked. There is nothing wrong with this account as far as it goes. The difficulty is that it does not go far enough. There are crucial differences between different types of network, which are often overlooked. Confusion is created when researchers fail to specify which type of network they are writing about.

Networks are everywhere. In *physics*, there are electrical circuits; in *civil engineering*, structures such as bridge trusses; in *information technology* (IT) there are telephones and the Internet; and in *geography* there are transport systems, such as motorways and railway systems. Agriculture and industry depend on distribution systems (pipelines, electricity grids) and disposal systems (drainage ditches, sewage systems). In *biology*, the brain is analysed as a network of neurons, and in *anthropology*, family networks are created and sustained through reproduction. *Economists* refer to networks of trade, investment and technology transfer when discussing international and inter-regional resource flows. *Sociologists* analyse social groups in terms of inter-personal networks, and use network effects to explain 'chain migration' flows, whilst *business strategists* analyse networks of strategic alliances between firms.

Ambiguities of the subject

The mathematical theory of networks is, rather misleadingly, termed the 'theory of graphs' (Biggs *et al.*, 1986; Diestel, 1997). Although graph theory claims to be general, it involves a number of simplifying assumptions that restrict its application in IBH.

When applying graph theory, different disciplines refer to the same concepts by different names. The members of a network are variously referred to as elements, nodes, vertices or points, while the connections between them are referred to as linkages, edges, paths and so on. Ambiguities exist even within a single discipline. In IBH the term 'network' is used in several ways (see Thompson, 2003). All four of the concepts described below have been applied to export-orientated industrial districts based on flexible specialization, but it is not always clear, in any given instance, which type of network a writer has in mind:

- A *'network' as a distinctive organizational form, intermediate between firm and market*. This type of network comprises a web of long-term co-operative relationships between firms. It is distinctive because the relationship between the firms is not authoritarian, like an employment relationship, and differs from a spot-market relationship because it involves a long-term commitment.
- A *'network firm' as a set of quasi-autonomous subsidiaries*. Japanese *keiretsu* and Italian business groups are often described in these terms. This type of network involves a small central locus of authority – namely an investor or group of investors who use a set of holding companies to control a range of businesses in which independent minority investors may also be involved. In mainstream IB literature the 'network firm' combines the global vision of an influential headquarters with the flexibility of autonomous subsidiaries.

- *A 'local business network' which involves key actors, such as bankers, entrepreneurs and government officials, who co-ordinate activities informally within an economic region or urban centre.* Unlike the previous cases, the network involves a mixture of organizations of different types. The relationships are used to finance strategic investments in local public goods, such as training colleges or dock improvements, whose benefits accrue to businesses in general rather than to any single business in particular.
- *A 'network industry', such as transport, water, energy and other utilities.* Network industries typically sink large amounts of capital into specialized infrastructure which links different locations and facilitates the movement of resources between them (Foreman-Peck and Milward, 1994). The network refers to the spatial linkages and the hubs at which they meet. Connecting an industrial district to a long-distance transport network is often crucial in promoting its export trade.

There is a tension between these specific connotations of a network, and the generality of the underlying concept. It can be argued, for example, that firms and markets are not alternatives to networks, but simply special types of network – the firm being a relatively rigid and hierarchical network, and a market a flat and flexible one. In this view, almost everything is a network, and so it is fruitless to argue about what is a true network and what is not. The research question is not so much, 'Is it a network?' as 'What type of network is it?' The key to understanding networks is to have a scheme by which to classify them.

Role of networks in the co-ordination of economic activity

In some disciplines, such as geography and sociology, networks are of intrinsic interest (see, for example, Grabher and Powell, 2004). In disciplines such as IBH, however, interest in networks is more instrumental: networks are studied because they help to co-ordinate IB activity.

Some sociologists have suggested that networks are created mainly because people like to belong to them. Intrinsic emotional benefits are important reasons for belonging to small and cosy groups, like a happy family. But not all networks are a pleasure to belong to; some professional networks can be very competitive, for example, and, far from welcoming new members, act more like a clique or a cartel. People still seek entry, however, because of the economic advantage that can be obtained (Casson and Della Giusta, 2007). If emotional benefits were the only ones that people derived from networks, it seems likely that networks would be much less common that they are.

Taking an instrumental view of networks helps to explain why there are so many different varieties in practice. Different network structures are best adapted to co-ordinating different types of economic activity. If emotional benefits were the only reward, it is likely that networks would be much more homogeneous: in particular, they would be much smaller and friendlier than many of them in fact are. To explain why network structures vary, it is necessary to recognize that different types of network co-ordinate different types of activity.

The second section of this chapter sets out a basic typology of networks, while the third section reviews the basic concepts of network theory, with an emphasis on network configuration. It draws heavily on the elementary theory of graphs, adapting the theory to the needs of historical research. The fourth section examines the different types of relationship from which a network can be formed, and re-examines some classic network issues in IBH. The fifth section examines the interplay between social and physical networks, while the final section summarizes implications for future research.

A simple typology of networks

Some basic distinctions

A *physical* network – for example, a road or river system – connects natural features, buildings and plants (Haggett and Chorley, 1969), while a *social* network connects people. In social networks, *social distance* is more relevant than *Euclidean distance*. An individual's social network may be summarized by the names in their address book, but it would be a mistake to suppose that those who live furthest away are contacted less frequently. Social distance may be expressed using a metric of communication costs, provided that these costs include not only the cost of a letter or telephone call, but also the costs of overcoming linguistic and cultural barriers.

Networks typically involve both *stocks* and *flows*. The stock comprises network *infrastructure*, while the flow comprises *traffic*. Social networks rely on an invisible infrastructure of shared languages and values, which builds reputations and generates trust; this infrastructure supports a flow of information between the members of the network. Physical networks are important in IBH for sustaining trade, while social networks are important for sustaining technology transfer, marketing and managerial communication.

Investment in networks

Another important distinction is between *natural* networks and networks *engineered* by human agency. Both physical and social networks can be engineered: thus a canal is an analogue of a natural river, while a club is an analogue of a biological family. Investment in engineering a network is intrinsically economic: it requires some person – or a group of people – to incur substantial present costs in anticipation of future benefits. This applies whether the objectives of the network are commercial or social. In general, engineering major networks requires entrepreneurship and leadership of a high order.

Engineered networks are typically embedded in natural networks: thus canals developed from cuts made in navigable rivers; and in the social sphere, members of clubs and communities may be also embedded in extended families.

Any given network is almost invariably part of a wider system. Connecting elements from different networks creates a 'bridge' between the networks. The importance of an element within any given network, such as that between local businesses, often derives from the number of external linkages that the element possesses; for example, the number of contacts a local business has in the nearest metropolis.

The only network that is not part of a wider system is the global network that encompasses the totality of all networks: it is the network that links every person, every resource and every location, directly or indirectly, to every other one. Every other network is a subset of this encompassing network. It is necessary to base analysis on subsidiary networks because this encompassing network is so complex that, while it can be analysed at a high level of aggregation, it is too large to analyse fully at a disaggregated level. It must be recognized that every subsidiary network selected for study is therefore an 'open system', which connects to the rest of the global network at various points.

The interdependence of physical and social networks

Social networks are used to manage flows through physical networks, and to co-ordinate strategic investments in them. Conversely, social networks require supporting services supplied by physical networks – for example, transport to and from meetings organized by a club. Geographers often study physical networks in isolation from social networks, whilst sociologists often study social networks in isolation from physical ones. Such partial perspectives provide a distorted picture of networks, and can result in misleading conclusions.

Physical networks involving flows of goods and services emerge because the *division of labour* leads individuals to specialize in particular tasks. A single complex task is broken down into a set of simpler tasks, each performed by a different person. Different elements of the physical network are created by this differentiation of tasks, different tasks are linked by flows of goods and services, and specialized hubs emerge where traffic flows converge. These hubs provide flexibility by allowing traffic to be switched from one route to another.

A social network is created to co-ordinate the actions of the people who have been assigned to different tasks, by improving communication between them. Social networks have their own divisions of labour too. Entrepreneurs and leaders act as information hubs. Individual consumers go to entrepreneurs to buy their goods, relying on the entrepreneurs to procure the goods on their behalf from the producer, who is the ultimate source of supply. Individuals who need to make contact with other individuals may go to a leader and ask for an introduction to be arranged. The leader may expect the individual to join the group, and possibly pay a membership fee, in return for receiving this service.

Competing hubs

In a private enterprise economy, entrepreneurs compete with each other for custom. Similarly, leaders of rival groups compete for members, and to gain influence for their views. As a result, both physical and social networks develop a multiplicity of competing hubs. Ordinary members use these hubs as gateways to the rest of the network. In effect, relationships between ordinary members of the network are mediated by the entrepreneurs from whom they buy, the leaders of the groups to which they belong, and the hubs through which they travel. While individuals also have direct connections to other individuals, the number of such direct connections is very small compared to the number of people to whom they are connected indirectly through the hubs. This would make the hubs extremely powerful if it were not for the competition between them.

Basic network theory: configuring connections

Connectivity and configuration

The defining feature of a network is *connection*. A set of *elements* connected to each other form a *network*. Every pair of elements belonging to a network is connected, either *directly* or *indirectly*. Indirect connections are effected through other elements of the network.

From an economic and social perspective, there are four key aspects of networks:

- *size*, as measured by the number of elements that belong to the network;
- *diversity*, as measured by the number of different *types* of element that belong to the network;
- the types of *relationship* that connect the members; and
- the *configuration* of the network, which describes the pattern in which the different elements are connected.

IBH historians have discussed relationships in considerable detail, but have said surprisingly little about size and diversity; that is, about the characteristics of the elements that are connected. However, the main deficiency in IBH is the lack of attention to configuration, and so this issue is the main focus of this chapter.

The exclusion of configuration is difficult to justify, given that it features significantly in the geographical literature. However, much of this research is focused on physical networks, while research on social networks tends to emphasize relationships. Only a small number of writers on social networks, such as Burt (1992), Leibenstein (1978) and Wasserman and Faust (1995), have integrated the analysis of relationships with the analysis of configurations.

There are many different ways of connecting a given set of elements. The configuration of a network is defined by the set of direct pairwise linkages between its elements. As the number of elements increases, the number of different ways in which elements can be connected increases dramatically. Network analysis is bedevilled by the complexity created by this 'combinatorial explosion'.

Complexity can be reduced by focusing on a small number of standard configurations, such as the hubs, webs and branches described below. These standard configurations can be combined in modular form to create large networks from sets of smaller networks – for example, a web of hubs, or a hub of webs.

Simple examples of standard configurations: linear and circular networks

Linear networks are widely used in transport systems – such as stations along a railway line – but are little used in social networks because they deliver poor service to the elements at the end of the line. The terminal elements are connected only directly to one other element, and connected to each other only by a path through all the other elements in the network. Repeated intermediation distorts communication in a social network but only slows down communication in a transport system, so a linear network is better suited to transport than to social interaction.

There may be more than one path between two elements. When travelling along a ring road, for example there is always an alternative route – for example, it is possible to get from east to west via either north or south. Where there are alternative routes, round trips are also possible, by going out along one route and back along the alternative. Alternative routes provide an element of redundancy in a network. While 'redundancy' sounds wasteful, it imparts flexibility to a network: if one linkage breaks, another can be used instead. In a social network, for example, if A wants to get in touch with B, they may have a choice of being introduced to B by either C or D. If either C or D falls ill, they can use the other person instead. Trading off cost and reliability of performance determines the optimal degree of redundancy in a network.

Hubs

Hubs are points at which three or more linkages converge: they act as consolidation and distribution centres for the traffic over the network. Hubs are often connected to other hubs by trunk connections which carry high-volume traffic (Watts, 2003). Unlike a ring, where all the elements are connected to two other elements, the ordinary members of a simple hub system are connected only to the hub. The great advantage of a simple hub configuration is that each element is not only connected directly to the hub element, but is also connected indirectly to every other element by a path comprising just one intermediate element – the hub itself.

The power of a hub can be measured by the proportion of through traffic it handles in proportion to the amount of traffic originating or terminating at the hub itself. When every linkage in a network carries the same amount of traffic, the power of a hub is proportional to the number of linkages it possesses. With n elements, including a solitary hub, and two-way flow of traffic x between each pair of elements, the total traffic through the hub will be $(n-1)$ $(n-2)x/2$. The traffic originating from, or destined for, the hub will be $(n-1)x$, and so the power of the hub will be the ratio of the first term to the second – namely $(n-2)/2$.

Webs

A weakness of the hub configuration is that there is no redundancy. A failure in any link will completely disconnect one of the elements from the network, and a failure of the hub itself is fatal. A natural solution here is to use more than one hub. In the limiting case, every element becomes a hub. This creates a web configuration, in which every element is connected directly to every other.

It is often said that networks afford significant economies of scale, but these economies are in fact attributable to hubs. In a web, where every element is connected directly to every other, the number of linkages, $n(n-1)/2$, is equal to the number of connections achieved, and so there is no saving in linkages as the number of elements in the network increases. On the other hand, the number of linkages in a corresponding hub is only $n-1$, and so network economies increase without limit when a hub configuration is adopted. The difference between the hub and the web becomes more pronounced as the number of elements increases, as moving from a hub to a web increases the number of linkages by a factor $n/2$.

Branch configurations

Hubs are prone to congestion. A hub carries both a physical burden of handling passengers and freight, and an information burden created because traffic arriving from each direction has to be switched on to the correct outward route. Satellite hubs can be created to share the burden. This leads to a pyramid (or 'branch') configuration.

Satellites hubs can be connected up to each other, creating a 'bypass' around the main hub – for example, most large cities are bypassed by roads which carry through traffic, leaving the city roads free to accommodate the substantial amount of traffic originating from or terminating in the city. Within organizations, the same principle encourages horizontal networking between senior managers, which prevents the chief executive from being bothered with issues that senior managers can resolve between them.

Relationships

A typology of relationships

There are many different types of relationship that can connect up networks. Failure to distinguish different types of relationship can cause serious confusion in the analysis of networks. Social relationships play a prominent part in institutional theory (North, 1981; Williamson, 1985), but they are usually discussed in terms of deviations from the market norm rather than as subjects in their own right. Granovetter (1985) provides the most sophisticated discussion of social relations in a network context, and the remarks below may be construed as a development and extension of his work.

It is sometimes assumed that relationships within social networks are symmetrical, but this is far from being the case. Social networks inside firms, for example, are often based on authority: the employment contract stipulates that a worker takes orders from his/her manager, and that a manager takes orders from the owners of the firm. Authority relationships may also be informal – as between a parent and child, for example. Formal and informal relationships may coexist, as when a formal contract of employment is supplemented by a mutual understanding between employer and employee.

Inward- and outward-looking social networks

Social relationships involve reciprocal obligations between parties. The obligations may be strictly mutual, so that the relationship is symmetrical, but even if the relationship is asymmetrical, the obligations need to be compatible. For example, if an employee has an obligation to carry out orders, then the employer has an obligation to give orders which the employee is able to implement.

Obligations are often specific rather than general – for example, they apply only to family and friends. Members of an 'inward-looking' network recognize only obligations to fellow members of the group, while members of an 'outward-looking' network accept obligations to the public at large.

Some networks emphasize uniformity, so that all members incur the same set of obligations. Uniformity is normally required in an outward-looking network that seeks to maintain the value of its external reputation, as with a professional association, where the public can be assured that any member of the association will maintain certain standards of behaviour. Members of reputable outward-looking networks are well-placed to act as intermediates in trade, because customers from other networks can be confident that they will not be cheated.

In an inward-looking group that lacks external reputation, personal knowledge of fellow members is important, in order to predict how they will behave. Such groups are good at providing mutual support for members, but not for developing people who can play a prominent role in trade.

For a person to know whether they can trust someone with whom they plan to trade, it is useful to know about the social networks to which they belong. It is prudent to trade with someone who either belongs to a reputable, outward-looking group, or to the same inward-looking group as themselves.

It is also useful to belong to a network whose members have contacts with many other networks, enabling them to check on other people through their own contacts. This is particularly important for entrepreneurs. Products in international trade are often bought and resold several times before they reach the consumer, which requires extensive trade between entrepreneurs. Given the high value of wholesale transactions, and the difficulty of enforcing international contracts through the law, trust is a crucial factor in international trade. Elite networks operating at the international level are therefore extremely useful, allowing entrepreneurs to facilitate international trade. Their success depends on each of the members having an extensive range of contacts to place at the disposal of other members.

Confidentiality and collusion in social networks

Most writers on networks assume that, if cost considerations are ignored, then more linkages are always better, because this will shorten the paths between some pairs of elements, and thereby reduce overall communication costs. However, in practice, many people try hard to avoid communicating with other people. One reason is simply that communication is time-consuming, and therefore costly.

There are numerous people we walk past every day in the street or at work to whom we do not stop and talk, because we are hurrying to meetings with people to whom we do wish to talk. Although a link to the people we walk past has been created by chance, we do not wish to take advantage of the opportunity. Even people we already know may be avoided if we do not trust them, for there is little point in talking to someone whom you do not trust, since you cannot believe what they say.

Other people are positively dangerous. An entrepreneur will not wish to communicate with a competitor because she does not wish the competitor to know his/her prices, because customers could be stolen away by the competitor quoting a marginally lower price.

Diversity of elements

Many networks involve different types of element. While some of the consequences of diversity are recognized within the literature, there is no systematic treatment of the subject. Within any network composed of different types of element, it is useful to distinguish linkages between elements of the *same* type from linkages between elements of a *different* type. Within a trading network, for example, a set of linkages between consumers and producers exemplifies a 'vertical' network, in which the flow of communication follows the same path as the flow of the product, while a network of producers – such as a cartel – and a network of consumers – such as a purchasing co-operative – both exemplify a 'horizontal' network between people operating at the same stage of product flow. It is often claimed that vertical linkages strengthen competition and increase trade, while horizontal linkages are anti-competitive and inhibit trade, but this is not always the case. The competitive implications depend on the interaction between the configuration of the network and the type of relationship involved; for example, a 'best practice' club involving producers, and price comparisons made by consumers through casual meetings can both improve the competitive performance of a trading network.

Transport and communication networks in the international business system

The demand for personal transport derived from social networking

There are close links between social and physical networks, with the flow of information through the former being used to co-ordinate the flow of resources through the latter. The links between social and physical networks become even closer when the spatial dimension of economic activity is examined in more detail.

There is an obvious stock/flow connection between product flow and physical transport infrastructure, and between information flow and physical communications infrastructure. It is not so obvious, however, that there is also an important connection between information flow and physical transport infrastructure. This connection arises because there is a crucial difference between remote and face-to-face communication: while the former creates a demand for communications infrastructure, the latter creates a demand for transport infrastructure. We cannot all live 'next door' to everyone else, so face-to-face communication creates a demand for travel to meetings.

When a group of just two or three plan to meet, people may take it in turn to act as host, but when a significant number of people need to meet, a central location will normally be used. A specialized central location reduces overall travelling distance, while large meetings economize on the use of time, as it is possible to meet many people by making just one journey.

Attending large meetings is also an efficient way of obtaining introductions: each person cannot only be introduced, but can also introduce others. The structure of the meeting is important in this respect. People need to be able to circulate so they can be paired up with appropriate others. Break-out areas in which people can hold confidential one-to-one discussions are also useful when the function of a meeting is to help broker business deals.

Efficiency of communication is increased if different meetings take place at the same central place, to enable people to attend several meetings during the same trip. Different meetings of interest to the same groups of people can be scheduled to run in sequence, as with the annual conferences of related professional associations.

Other meetings are in continuous operation. A shopping centre, for example, may be construed as a continuous open meeting, where people can come and go as they please. People who are attending scheduled meetings can 'pop out to the shops' at their convenience. Retailing is a prominent activity at many of the hubs where people meet. Historically, abbeys and castles attracted retailers, especially on saints' days, while many of today's major retail centres originally developed around ports or centres of government.

Emergence of personal transport hubs through economies of agglomeration

Retailing affords economies of agglomeration to consumers. Where different retailers stock different types of product, the consumer can collect an entire 'basket' of different goods on a single trip. Where different retailers stock different varieties of the same product, the customer can assess the design and quality of different varieties on the spot. Where different retailers stock the same product, customers can compare prices. In each case the agglomeration of retailers reduces the marginal cost of a customer's search.

Retailers supplying complementary goods have a direct incentive to locate close together. If one shop is already selling outerwear, for example, then a shop that sets up next door selling underwear can anticipate a substantial 'passing trade'. It is not so clear, though, why competing retailers would locate together. One reason put forward for customers refusing to buy from a local monopolist is that they believe they will be cheated. This reflects a lack of trust in society. Thus retailers who locate together gain credibility: they acquire a small share of a large market instead of the entire share of the very small market they would otherwise enjoy.

Another explanation relates to innovation. An effective way to advertise a new product is to display it adjacent to its closest competitor. Customers can be 'intercepted' on their way to their usual source of supply. The 'market test' may well put one of the suppliers out of business: if the new product is successful, then the established retailer may quit, while if it fails then the innovator will quit. An established retailer defeated by an innovation may retire to a more remote location where they can monopolize a small market with their traditional product.

The link between innovation and agglomeration explains why a market may be regarded as a 'self-organizing' network. Volatility in the environment continually creates new consumer problems, and a consequent demand for new products to solve them. At the same time, the social accumulation of technological knowledge allows new types of product to be developed. But an innovation is only viable if it can find a market, and its market is to be found where its closest competitor is sold. To make as much profit as early as possible, a confident innovator will head for the largest market (possibly after 'proving' the product in a smaller market first), which is where the contest between the new product and the old product will be played out.

The continuous influx of new products increases both the novelty and the diversity of the products available at a major hub. The greater intensity of competition means that older, obsolete products will be expelled from large markets before they are expelled from smaller ones. The larger the market, therefore, the greater the diversity and the lower the average age of the product.

On the other hand, the risks faced by consumers are greater in a large market, because a higher proportion of the products will be unproven. A large market will therefore attract buyers who are confident of their ability to judge design and quality, and who value novelty for its own sake, while smaller markets will retain the custom of less confident people, and those who prefer proven traditional designs. Optimal innovation strategy therefore explains both the capacity of the market system to renew itself continually by updating its product range, and the concentration of novelty in the largest markets.

Varieties of transport hub

Meeting points and major markets constitute information hubs. People visit these centres specifically to meet other people. Shopping sustains profitable production, while meetings support the innovation process: researchers 'network' at conferences, entrepreneurs meet venture capitalists at elite gala events, and inter-firm alliances are planned at trade fairs.

With so many visitors to the hub, accommodation, catering and entertainment facilities are required. The ease of access and variety of services available at the destination attract tourists. A *visitor hub* of this type is, in principle, quite distinct from the *transit hub*, such as a railway junction or airport hub, at which people change from one trunk route, or transport mode, to another. People travel *through* a transit hub in a particular direction, whereas they travel *to* and *from* a visitor hub as part of a return journey.

The essence of a transit hub is that a number of connecting trunk lines converge at the same point. Some traffic can be switched from one route to another without stopping – for example, express trains at country railway junctions and motor traffic at motorway intersections – but in other cases a stop is required so that a connection can be made with another route. It is when traffic has to stop that it may 'stop over' rather than proceed on its way at the first available opportunity.

There is little point in stopping over at a pure transit hub, as there are no major services to attract the visitor, but a visitor hub, on the other hand, can attract stopovers if it can also be used as a transit hub. In order to act as a transit hub, however, it needs to occupy an appropriate location on the transport network, at the intersection of important routes.

Combining the role of transit hub and visitor hub can lead to congestion, however. In the late twentieth century, visitor attractions tended to concentrate in the centres of cities, while transit hubs moved to airports and motorway junctions on the periphery. In the global economy of the twenty-first century, competition between 'world cities' is based on finding an efficient way of combining the roles of transit hub and visitor hub.

Conclusion

Networks have stimulated much interest in IBH since the mid-1990s. Part of this is because of their ideological significance. They have been hailed as an alternative to large, impersonal organizations such as a rigidly hierarchical multinational firm or the state. Indeed, it has been suggested that the modern capitalist system took a 'wrong turning' around the beginning of the twentieth century, when the large managerial corporation superseded the networks of flexible specialization that prevailed in the industrial districts of the time (Piore and Sabel, 1978).

Networks are inherently complex, but this does not mean that they cannot be properly understood. The structure of a network is governed by four main factors:

- the size of the network, as measured by the number of elements;
- the membership of the network, as reflected in the types of element that belong to it, and the extent to which different types are mixed;
- the types of relationship between members, which reflect the roles that they play; and

- the configuration of the network, which describes the pattern by which the different elements are connected.

Recent analysis of social networks has been dominated by the study of relationships, and in particular by the issue of trust. This has distracted attention from the issue of configuration. Configuration is an important influence on the cost of operating a network. It is the major focus of graph theory, and has received much attention in research on physical networks, but has been ignored by most writers on IBH.

This chapter has outlined the structure of a positive theory of networks, which explains why certain types of network are particularly common in certain situations. The division of labour provides the rationale for many physical networks. Different co-ordination requirements are best satisfied by different network structures. Hence the nature of the division of labour determines the pattern of co-ordination required, which in turn determines the most appropriate network structure. If co-ordination is efficiently organized, then the most efficient network structure will be the one that is used.

Long-distance trade, for example, is usually co-ordinated by for-profit entrepreneurs through inter-firm contracts, while the delivery of local social services is usually co-ordinated by non-profit leaders, who establish schools, hospitals, churches, sports clubs and community associations for this purpose.

Because each person in the economy belongs to many different networks, all the networks to which people belong are intertwined. Every network is connected, directly or indirectly, to every other network by multiple links. Thus every network is a subset of a single giant network that encompasses the entire global economy.

References

Biggs, Norman L., E. Keith Lloyd and Robert Wilson (1986) *Graph Theory, 1736–1936* (Oxford University Press).

Burt, Ronald S. (1992) *Structural Holes: The Social Structure of Competition* (Cambridge, Mass.: Harvard University Press).

Casson, Mark (1997) *Information and Organization* (Oxford: Clarendon Press).

Casson, Mark (2000) *Entrepreneurship and Leadership* (Cheltenham: Edward Elgar).

Casson, Mark and Marina Della Giusta (eds) (2007) *Economics of Networks* (Cheltenham: Edward Elgar).

Diestel, Reinhard (1997) *Graph Theory* (New York: Springer).

Doreian, Patrick and Frans N. Stokman (1997) *Evolution of Social Networks* (Amsterdam: Gordon & Breach).

Foreman-Peck, James S. and Robert Millward (1994) *Public and Private Ownership of Industry in Britain, 1820–1980* (Oxford: Clarendon Press).

Grabher, Gernot and Walter W. Powell (eds) (2004) *Networks* (Cheltenham: Edward Elgar).

Granovetter, Mark (1985) 'Economic Action and Social Structure: The Problem of Embeddedness', *American Journal of Sociology*, 91(3), 481–510.

Haggett, Peter and Richard J. Chorley (1969) *Network Analysis in Geography* (London: Edward Arnold).

Jones, Geoffrey G. (2000) *Merchants to Multinationals: British Trading Companies in the Nineteenth and Twentieth Centuries* (Oxford University Press).

Jones, Geoffrey G. and Franco Amatori (2003) *Business History around the World* (Cambridge University Press).

Knight, Frank H. (1935) *The Ethics of Competition and Other Essays* (London: Allen & Unwin).

Leibenstein, Harvey (1978) *General X-efficiency Theory and Economic Development* (New York: Oxford University Press).

North, Douglass C. (1981) *Structure and Change in Economic History* (New York: W. W. Norton).

Piore, Michael J. and Charles F. Sabel (1978) *The Second Industrial Divide: Possibilities for Prosperity* (New York: Basic Books).

Thompson, Grahame F. (2003) *Between Hierarchies and Markets: The Logic and Limits of Network Firms* (Oxford University Press).

Wasserman, S. and Faust, K. (1995) *Social Network Analysis: Methods and Applications* (Cambridge University Press).

Watts, Duncan J. (2003) *Six Degrees: The Science of a Connected Age* (London: William Heinemann).

Williamson, Oliver E. (1985) *Economic Institutions of Capitalism* (New York: Free Press).

2
Subsidiary Entrepreneurship Orientation: The ThermoSafe Case

Francesco Ciabuschi and Mats Forsgren

Introduction

This chapter deals with 'entrepreneurship orientation' at the subsidiary level. A reasonable assumption is that the basic factors behind entrepreneurship orientation are the same irrespective whether we are focusing on independent firms or on subsidiaries in MNCs. While the essential act of entrepreneurship is *new entry* (into markets and/or products), entrepreneurship orientation is contingent on the actor's autonomy, innovativeness, risk-taking propensity, proactiveness and competitive aggressiveness. These dimensions may vary independently of each other, but are all positively related to performance in terms of new entry. Through a case study we argue that specific conditions related to the MNC context must be included in a model of 'subsidiary entrepreneurship orientation'. Tentative results of the case analysis are that (i) the autonomy variable must be complemented by an analysis of the degree of visibility of the 'rules of the game' on which the perceived autonomy is based; (ii) the subsidiary's risk-taking propensity is dependent on the 'organizational risk' – that is, the subsidiary perception of the possibility of losing internal legitimacy as a consequence of the venture; and (iii) a subsidiary's entrepreneurship orientation is also contingent on the 'power struggle' within the MNC.

While much is known about entrepreneurship and entrepreneurial processes in general (Aldrich, 1999; Shane, 2000), the issue of entrepreneurship specifically in MNCs involves a number of relevant and urgent research questions. One is the attempts by MNCs to access a greater diversity of business opportunities by creating dual roles at the subunit level: (i) the management of ongoing business activities; and (ii) the identification and pursuit of new opportunities. This approach, sometimes called 'dispersed corporate entrepreneurship' (Birkinshaw, 1997), means that the entrepreneurial capability is not restricted to a separate unit within the MNC. It rests on the assumption that is possible to reach a balance between managerial responsibilities and entrepreneurial responsibilities at the individual level, so that the former does not 'drive out' the latter (Birkinshaw and Ridderstråle, 1999). The issue of under what circumstances achieving such a balance is possible is so far an under-researched area among international business scholars.

A related issue concerns the inherent tension between the role a subsidiary is expected to fulfil in the MNC's corporate strategy, and its role in the local business network in which it is embedded (Andersson and Forsgren, 1996). The overarching question implicit in these types of trade-off is to what extent subsidiaries can function as entrepreneurs within the MNC. This question branches out in turn into a number of issues. For example, urgent research questions are: the possibility of subsidiaries taking initiatives and getting support for them from headquarters (Birkinshaw, 1997; Goodall and Roberts, 2003); the role of the local environment for the subsidiary's ability to be entrepreneurial (by gaining access to resources outside its direct control); and the impact of a 'control gap' between the headquarters and the subsidiary on the subsidiary's entrepreneurial capability. A more general issue in this context is to what extent subsidiaries of MNCs display a greater degree of entrepreneurial orientation than that seen in subunits of a national firm (Yamin, 2002).

Entrepreneurship orientation at the subsidiary level

The essential act of entrepreneurship is *new entry*: by entering new or established markets with new or existing goods or services. This act, in turn, is contingent on the actor's *entrepreneurship orientation*, which basically consists of five dimensions: the actor's autonomy; innovativeness; risk-taking propensity; proactiveness; and competitive aggressiveness (Lumpkin and Dess, 1996).

In a MNC/subsidiary context, autonomy deals with the freedom the subsidiary has to take initiatives and decide about investment in new markets and/or new products/services; that is, the degree of formal decentralization. Innovativeness reflects the subsidiary management's willingness to try new entries, in terms of new technology or markets. Risk-taking propensity deals with the subsidiary management's readiness to venture into the unknown, committing a relatively large proportion of assets or borrowing heavily (Baird and Thomas, 1985). The latter two dimensions reflect the financial risk, while the first is related to a venture's business risk. Proactiveness emphasizes the importance of taking initiatives in the entrepreneurial process. A proactive subsidiary seizes opportunities by following a first-mover strategy and employs a forward-looking perspective that is accompanied by innovative venturing activity (Lumpkin and Dess, 1996). While proactiveness deals with seizing opportunities in order to shape the environment, competitive aggressiveness, deals with how the actor relates to its competitors. A subsidiary's competitive aggressiveness reflects its readiness to challenge its competitors to achieve entry or an improved position; that is, to outperform industry rivals in the marketplace.

It has been argued that these five dimensions are salient aspects of entrepreneurship orientation. They may vary independently of each other but are all positively related to performance in terms of entrepreneurship, that is, new entry (Lumpkin and Dess, 1996). It is also argued that this model is as relevant for a subunit in a large organization as it is for a small, independent firm. Expressed differently, entrepreneurship orientation, and therefore the propensity of new entry, is founded on some basic principles, which are the same irrespective of whether an independent firm or a subsidiary in an MNC is in focus. An MNC subsidiary's propensity to act as entrepreneur can be analysed exhaustively using these dimensions.

There are reasons, though, to question this conclusion. Through an illustrative case[1], presented below, we argue that there are specific conditions related to being a subsidiary in an MNC which should be considered, in order to build a model of a subsidiary's entrepreneurship orientation, and consequently of its propensity to take on a new entry behaviour.

The case of ThermoSafe

Svenska Cellulosa Aktiebolaget (SCA) growth in the USA

SCA is a Swedish-based MNC producing and selling absorbent hygiene products, packaging solutions and publication papers around the world. SCA has about 51,000 employees in some fifty countries, and its annual sales amount to more than €10 billion (in 2005). SCA is structured formally into six different divisions: SCA Packaging Europe, SCA Tissue Europe, SCA Personal Care, SCA Forest Products, SCA Americas and SCA Asia Pacific (see Figure 2.1).

Since 2001, SCA Americas' business, located mainly in the USA, has grown exponentially through acquisition (SCA bought twelve US firms between 2001 and 2004). At the time of writing, the SCA organization in the USA accounts for almost 10,000 employees, has about seventy manufacturing locations and is structured in three different subdivisions: SCA Tissue AFH North America, SCA Personal Care North America, and SCA Packaging North America (see Figure 2.1). SCA Packaging North America (NA) is further structured according to the three types of packaging solutions offered: protective, temperature assurance or high visibility. Each of the three business areas is connected to very specific acquisition that SCA made between 2001 and 2003. In the case of protective packaging, the milestone was the acquisition of Tuscarora in 2001; in the case of temperature assurance, the acquisition of four highly specialized firms – ISC Labs (2001), Polyfoam Packers (2002), Mid-Lands Chemicals (2002) and H&R Industries (2003); and for high-visibility packaging, the acquisition of Alloyd in 2003.

The four firms specializing in temperature assurance solutions, initially part of the protective packaging business under Tuscarora management, were

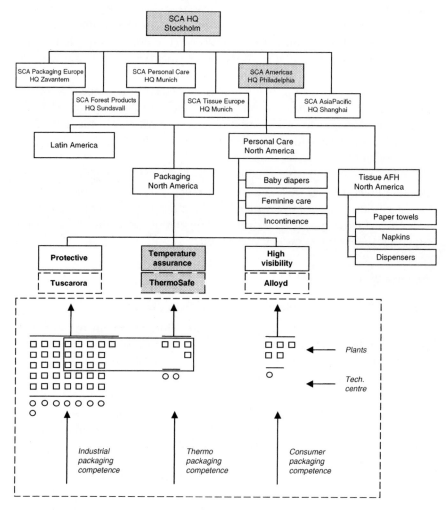

Figure 2.1   SCA organization in the USA (2005)

merged together in 2003, following the vision of Mr O'Leary – former owner of Tuscarora and president of SCA Packaging North America (2001–4), to form a specialized independent venture, 'ThermoSafe'.

ThermoSafe foundation

ThermoSafe is a provider of temperature assurance and testing solutions in North America. It offers products and services based on innovative packaging technologies including gels, insulation, containers and monitors, and it specializes in foam moulding, rigid PUR manufacture, gel-pack manufacture and rotational moulding. ThermoSafe sales are mainly within the USA, and amount to USD100 million, with an 18 per cent operating profit in 2004.

Although ThermoSafe is a relatively small investment for SCA (about USD60 million), its importance was recognized, as it provided the opportunity to enter a new packaging niche – that is, 'temperature assurance' packaging. Still, the temperature business is not new only to SCA but was quite new in absolute terms as the market itself is still emerging. In fact, the ThermoSafe operation contributed to shaping a recognized, specialized niche within the packaging industry.

Initially, Tuscarora was pivotal in the formation of ThermoSafe. Tuscarora was founded in 1962 in New Brighton (Pittsburgh, Penn.) and acquired by SCA in 2001 with the intention of expanding into the protective packaging business within the USA. Tuscarora was directly involved in the acquisition of ISC, Polyfoam and Mid-Lands, and until 2003 (the year of ThermoSafe's formation) these three firms were controlled by Tuscarora. At the time of writing, although ThermoSafe is an independent venture, Tuscarora is still involved, as part of ThermoSafe production takes place on demand at thirteen (of the forty-plus) Tuscarora plants.

The protective packaging company ISC Inc. was acquired in 2001 for USD10 million. The Arizona-based company produces high-performance insulated shipping containers for transporting temperature-sensitive products, including biopharmaceutical products, blood components and diagnostic samples. At the time of the acquisition, ISC, with two plants and one laboratory (in Phoenix, Arizona), generated annual sales of USD12 million. A second laboratory was established some time after the acquisition in Harleysville (Penn.), to better supply customers on the East Coast.

In 2002, SCA acquired Polyfoam Packers Corp. for USD22 million. This company also specializes in high-quality protective packaging for the shipment of similar temperature-sensitive products as ISC Inc. With production in Wheeling (Ill.) and Michigan City (Ind.), the company served customers nationwide in the US. At the time of the acquisition, Polyfoam had 225 employees and annual sales of about USD30 million. Shortly after the acquisition, the Wheeling plant was shut down.

Also in 2002, SCA acquired the protective packaging company Mid-Lands Chemical Company Inc., for USD16.8 million. Mid-Lands manufactures artificial ice and other refrigerants within the 'cold-chain' segment. The company's products are sold primarily to pharmaceuticals and diagnosis companies, but also to companies that manufacture and distribute food products that are sensitive to temperature. At the time of the acquisition, Mid-Lands, with six sites, had fifty employees and annual sales of approximately USD11 million. In 2005, none of these sites remained within Thermosafe.

In 2003, few months after the official establishment of ThermoSafe, H&R Industries Inc. was bought by SCA and added to its organization. H&R

manufactures a line of innovative and high-quality insulated fibreglass and polyethylene returnable container systems for storing and transporting temperature-sensitive frozen and perishable food products. H&R, with fifty employees, operated from two manufacturing facilities, one in Beecher (Ill.) and the other in Spartanburg (SC). The Spartanburg plant was disinvested some time after the acquisition.

Thermosafe's early development

Between 2003 (when ThermoSafe was founded) and 2005 there was substantial reorganization of the four acquired firms into ThermoSafe. The overall number of facilities decreased after the acquisition to nearly half, although a second laboratory was established. In particular all the Mid-Lands plants were disinvested. These disinvestments meant that many employees, including managers, were laid off, and the number of products manufactured by ThermoSafe decreased. All the decisions regarding dis/investment were taken by ThermoSafe without any interference from the larger group. The structural changes are summarized in Table 2.1.

Overall, according to the ThermoSafe top management, the acquisition and merger of the four companies forming ThermoSafe generated a positive effect in terms of sales, profit margins and market share, motivation of employees, corporate culture and production efficiency. However, as Ken Harris, Vice-President of SCA Packaging North America and responsible for ThermoSafe operations, explained: 'Although the overall satisfaction, many of the expectations connected to the acquisitions have not been fulfilled yet'. In particular, the integration process is judged by ThermoSafe management to be slow compared to what was established *ex-ante*, and some cultural problems emerged between parts during the process. One example of a cultural problem faced by ThermoSafe is that with Tuscarora. Tuscarora's key role in shaping the ThermoSafe organization created a kind of hierarchical relationship between the two, which was in contrast to the intention of having an independent business devoted to the temperature-assured packaging business, on the basis

Table 2.1   Changes in ThermoSafe's structure

	2003	Changes	2005
ISC	2 plants, 1 lab	+ 1 lab	2 plants, 2 labs
Polyfoam	2 plants	– 1 plant	1 plant
Mid-Lands	6 plants	– 6 plants	0 plants
H&R Industries	2 plants	– 1 plant	1 plant
Tuscarora	*13 plants**	–	*13 plants**

Note: *13 Tuscarora plants across the USA occasionally produce on demand for ThermoSafe.

of different customer bases, technology utilized, types of services, and marketing strategies. According to Mr Harris, the main problem ThermoSafe had to face during its early development, apart from the business challenge *per se*, was 'to win total independence from Tuscarora'.

As a result of the reorganization, the number of customers grew, while the number of suppliers decreased. In addition, ThermoSafe's relationships with customers are much closer compared to average packaging industry standards, and to the average relationships of each of the four firms acquired. For example, ThermoSafe have discussions, training sessions and meetings (for example, workshops or round-table discussions) with some customers on a regular basis. This helps to provide highly specific services (for example, in packaging design, testing and improvement). In particular, business interaction with large customers such as Merck and Wal-Mart has been increasing rapidly because, as ThermoSafe's Marketing Director commented, these big firms do not recognize ThermoSafe *per se* as their supplier, but rather ThermoSafe as part of SCA, a leading and trustworthy MNC with more than 50,000 employees around the world. However, at the same time, customers of this kind expected much more from ThermoSafe, precisely because ThermoSafe is identified with SCA. This means for example that: '*mistakes that could be forgiven to a small company, such as Polyfoam or Mid-Lands, are not forgiven now to ThermoSafe*' (Ken Harris).

From an organizational perspective ThermoSafe was formed as an independent business unit within SCA and fully responsible for running its own operations within the USA. Formally, ThermoSafe is part of SCA Packaging North America and answerable to the SCA Americas Division HQ located in Philadelphia. All its functions are independent of HQ or other units, though there are some occasional contacts with SCA Packaging HQ (Brussels) because of technology and patents matters. However, apart from this there is no overall co-operation with ThermoSafe's European units or corporate HQs.

According to ThermoSafe engineers, the company is unique within SCA for the methodology used in developing and finalizing customized solutions – for example, no other unit within SCA has testing facilities devoted to temperature assurance. ThermoSafe, though centrally recognized within SCA for its unique competence, has no official role assigned by HQ in connection with that (for example, as a centre of excellence), nor is any other unit making use of its competence. Moreover, ThermoSafe, being in a rather specific business niche, has market conditions that are unique compared the rest of the SCA units. For example, many of its customers and competitors are not potential customers (or competitors) of any other SCA, unit mainly because they are not buyers of packaging but only of thermo-protective solutions, which also explains why ThermoSafe is not in direct competition with any other of the company's units.

Within the USA there has recently been an attempt at co-operation between ThermoSafe, Tuscarora and Alloyd, in relation to the technical/product side. The intention is to cross-fertilize the different segments, but so far very little investment has been made. Apart from this, SCA Packaging North America's technical and product development centres are geographically separated and organized according to their focus and organizational membership – that is, the thermo packaging segment (ThermoSafe), industrial packaging (Tuscarora) and consumer packaging (Alloyd). The only form of co-operation is, as already mentioned, between ThermoSafe and Tuscarora related to production capacity.

To sum up, we quote Ken Harris's observation:

> We [ThermoSafe] were helped by SCA only in terms of financial resources, especially at the beginning. We tried to connect with the rest of SCA, but it did not work. Thus, we are not really integrated within the group and we are acting autonomously. We think of ourselves as highly entrepreneurial. We are working our way through the market to get legitimate within the company – once we win the market we have to get credit from the rest of the organization. We don't care to compete against Tuscarora or some other SCA unit that is against us, what matters is to succeed in the market. I don't know if that is the only possible path to take, but to us it seems the right one.

ThermoSafe's expansion plan

From a business perspective, ThermoSafe is currently aiming at expanding its operations into Europe. Such plans have been scheduled by ThermoSafe since its formation, but there have been organizational problems hindering them so far, in particular little overall communication and some cultural problems. Regarding the latter, problems emerged in particular because of the other units' perception of ThermoSafe as a different kind of unit with a different strategy and business. In the words of Kevin Grogan, ThermoSafe Director: 'Our problem is that the expansion into Europe is critical for our survival and success, but all our efforts have always crashed on the rocks of SCA European politics.'

This situation did improve when the responsible of SCA Packaging North America changed in mid-2005. The new president recognized the necessity for ThermoSafe to expand overseas, and the opportunities this would also provide for the rest of SCA's business. He admitted, however, that there are too many organizational obstacles in ThermoSafe's international expansion path, and the only way to take ThermoSafe into Europe was through an acquisition. Immediately, ThermoSafe identified a UK-based firm as a suitable target for acquisition.

ThermoSafe management believes, however, that they still are very late in expanding into Europe, and their struggles are not yet over. In particular, because ThermoSafe's decision to buy the overseas firm coincided with the decision by SCA HQ (Stockholm) to stop temporarily all acquisitions planned within the group. To get around this problem, ThermoSafe invested in a minority share (25 per cent) instead of acquiring the whole firm. In this way, the operation appeared to be an investment rather than an acquisition. ThermoSafe at the same time wrote an agreement to buy the remaining 75 per cent after three years, with the plan and hope that as soon as SCA HQ would allow acquisitions to take place again, they could finalize the acquisition.

According to Kevin Grogan, the primary risk related to this operation is that if the HQs do not revoke the block on acquisitions during the next couple of years, ThermoSafe would not only lose the financial and time investment made, but also a good partner and the possibility of penetrating Europe. In such an event there is also the risk that the UK firm would independently exploit the competences and technology that are transferred from ThermoSafe, which would represent not only a loss but also the birth of a new strong competitor.

The overall expansion plan foresees by the end of 2006 a network of five facilities located in different European countries, and two laboratories for testing. In 2007, as well as the planned further development of the business into Europe, ThermoSafe is also considering internationalization into Asia. With the intention of always being one step ahead of competitors, ThermoSafe considers entering Asia, and in particular China, as the last step in developing a potentially global business. As described by Kevin Grogan:

> It has always been our vision to run the ThermoSafe business from Chicago at a global level. SCA HQ did not say anything, in favour or against. So still today we believe in our ideas but we don't know if HQ will prefer us to run the business globally or if they would like us only to transfer competence to the other regional divisions. We are left alone and we have no information from the centre, so we are making our bets.

Central to ThermoSafe's plan for a global reach is, however, a stable growth and increasing success within the USA itself. According to Ken Harris, to sustain global development, ThermoSafe has to be doing extremely well back home. Financial, human and technological resources within ThermoSafe have to be enough to permit the company to be independent of SCA in its internationalization.

ThermoSafe development struggle – problems and risks

According to Ken Harris, the expansion into Europe represents a big risk for ThermoSafe, not only in economic and business terms but also from an organizational point of view, since there is no guarantee, as mentioned earlier, that ThermoSafe will be left to run the operations overseas. Ken Harris commented further: '*Probably the most risky thing that we as ThermoSafe have ever done is to move somewhat against the SCA organization and take our first step into Europe through an acquisition by having in mind the vision to make ThermoSafe a global venture.'*

Since the beginning, ThermoSafe's approach to the market, which is unique within SCA Packaging, caused problems for the acceptance of ThermoSafe within the organization. However, ThermoSafe is defending its own business model, since it has been repeatedly proved to be successful and no interference has been forthcoming from HQ on ThermoSafe's domestic matters. But this situation degenerated to the point where ThermoSafe has been pointed out by other units as a threat to their businesses. This has been evident, for example, in the conflicts between ThermoSafe and Tuscarora, and with certain European units. Nevertheless, as Ken Harris further commented: 'the difference between the US situation and the European one is that the latter is 100 times worse than in the US as the SCA organization in Europe is much larger, much more rooted into the company heritage, and the market is much more segmented than it is in the US.'

The problem overseas seems to be ingrained at national subsidiary level; the Brussels HQ is trying to support ThermoSafe's plans but at the same time they do not want to alter the existing equilibrium in each of the countries, and are afraid of creating problems in the existing organization. ThermoSafe trusted Brussels to channel them into Europe, but they were never in fact committed to that. As a ThermoSafe manager commented: 'People in Brussels were good listeners and they have been promising support, but during the past years they never actually did anything concrete and I believe that they did not have any clear plan for ThermoSafe more that agreeing to the idea that it was good for us to come to Europe.'

As for the Stockholm HQ, after initial involvement during the acquisition phase of the original companies that formed ThermoSafe, they were never directly involved again with issues related to ThermoSafe. ThermoSafe's plans for international expansion have also been delayed by the fact that HQs, at both divisional and corporate levels, have not put ThermoSafe at the top of their priory list, since their business is doing well and SCA has had to face other more urgent and problematic issues.

The general belief within ThermoSafe is that once it has grown larger by collecting different market successes, then they might have the strength to become more visible within SCA without the risk of immediately being stepped on by other units.

Aiming at gaining more legitimacy ThermoSafe has been looking at SCA Hygiene's operating methods, which are reputed to be more similar to theirs than those of SCA Packaging. In fact, SCA Hygiene is running its business in a consumer-orientated way, with a pan-European structure, with key accounts, brands and so on. The packaging business is very much a local business and what ThermoSafe is trying to do has never been done before. ThermoSafe aims to develop the first pan-European operation within packaging.

Within SCA Packaging's current European organization two different types of unit and business are being run. On one side there are specialized units, which are the sole producers of a particular kind of packaging sold all over Europe, and on the other there are traditional types of packaging units (box plants or converting plants). These are more numerous compared to the previous category, are very local (customers are no more than 250 km away from the plant), and they produce and sell standard types of packaging – for example, corrugated boxes. The ThermoSafe approach (and plan for Europe) is different from either of these ways of organizing and operating as it is based somewhere between the two types. In fact, ThermoSafe foresees for its business a co-ordinated effort of several units from different regions acting with speciality products and customized solutions of high value added (including testing services), which should deliver the same products and standards throughout Europe by having sales and marketing functions in Europe centrally coordinated. And as a ThermoSafe engineer commented, this is something that SCA Packaging is not used to doing.

Some of the Packaging units in Europe are already serving potential ThermoSafe customers with somewhat simple solutions resembling those of ThermoSafe. Basically, these units are offering some extra components with their ordinary packaging solutions. But this is not recognized as a different kind of business, and is considered merely to be part of standard packaging solutions. There are only a few SCA units that are already providing some kind of thermo-protective packaging in some locations around Europe (four units have been identified by ThermoSafe analysis) and so far they have not tried to develop the business in the way ThermoSafe has. As Kevin Grogan commented: 'SCA does not see that it already has a sort of raw ThermoSafe business in Europe which could be easily pulled together, shaped up and rendered as an official business segment – a competent sales and marketing organization could favour customers by offering better products and services across the whole of Europe.'

That is why, recently, ThermoSafe managers had a meeting with some local subsidiary managers of SCA Packaging in Europe and tried to promote the idea that a thermo business within SCA Europe should be formed, and that ThermoSafe intended to be the promoter of such an initiative. Kevin Grogan describes local European SCA managers' reactions as being mixed: 'Some managers, although agreeing with the idea, did not want to be involved, and others, although interested, were convinced that there was not enough business in the region, and finally there were those not interested at all.' He commented further: 'European subsidiaries seem to have their own issues and they don't tell us what these issues are, and that is the reason why we find a barrier to our overseas plans.' As some Thermosafe managers commented, a key point is that while there are problems between ThermoSafe and Tuscarora, ThermoSafe knows well Tuscarora capabilities and activities, and has access to Tuscarora's manufacturing facilities when needed. This is not the case with the European units. Until there is no open communication with the European units much is precluded.

Analysis and concluding discussion

A look at the case above seems to indicate that ThermoSafe fulfils quite well the five requirements listed earlier, reflecting a high degree of entrepreneurship orientation. First, it enjoys a high degree of *autonomy* in relation to the corporate HQ in Sweden, which is in line with how Swedish MNCs traditionally manage their foreign subsidiaries. ThermoSafe's autonomy also reflects the limited knowledge at the corporate HQ of ThermoSafe's business, and the fact that the firms from which ThermoSafe was created are all relatively newly-acquired ones.

Second, ThermoSafe demonstrates a profound willingness to enter a new market – in this case, Europe. This has been part of the subsidiary's business plan ever since the formation of the company. So, in this sense, the subsidiary's orientation is highly *innovative*. Third, ThermoSafe's management realizes that the plans for growth into Europe mean a substantial business risk, because, for example, of the difficulties in evaluating the market potential and the reactions of competition. The subsidiary also realizes that entry into Europe cannot be achieved through a stepwise procedure but has to be carried out by an acquisition of a relatively large European firm. The financial risk is therefore substantial. Despite the fact that ThermoSafe's management is aware of the relatively high business and financial risks, they are prepared to invest in the project to enter the European market. Therefore, ThermoSafe's attitudes reflect a relatively high *risk-taking propensity*.

Fourth, ThermoSafe is eager to get ahead of competitors and therefore be one of the first in the market, and is concerned with the time it takes to realize the plans. This is one reason why acquisition is preferred as an entry mode, because this shortens the time it takes to establish a position and shape the market environment. ThermoSafe therefore demonstrates a high degree of *proactiveness*. Finally, the fact that ThermoSafe also has the ambition to expand not only into Europe but also to Asia is an indication of its *competitive aggressiveness*.

To sum up, we can conclude that on all the five dimensions of entrepreneurial orientation, ThermoSafe scores high rather than low. It is therefore reasonable to expect that the subsidiary would be a case of high entrepreneurial activity. ThermoSafe's management also think of themselves as 'highly entrepreneurial'. However, the entry into Europe seems to be a slow and cumbersome process which does not seem to be totally in line with what one could expect, considering the high entrepreneurial orientation of ThermoSafe's management. There are seemingly barriers to carrying out the project. Therefore, it is reasonable to question whether there are other factors, related to the fact that we are dealing with a subsidiary in an MNC rather than an independent firm, that are not covered by the five requirements listed above? Are there other dimensions that should be considered in order to assess the entrepreneurial orientation of a subsidiary in a multinational corporation?

The case produces some indications in that direction. Take first the autonomy dimension. It is clear that ThermoSafe enjoys a very high degree of autonomy, and in that sense has the possibility of using its own initiative, as well as having the possibility, because of its relative financial strength, of deciding about investments without formal support from the corporate HQ. In that sense, there are few barriers to entrepreneurship at the subsidiary level. But it is crucial to estimate *why* the subsidiary enjoys a high degree of autonomy, and how it affects the daily life of the subsidiary. The high degree of autonomy is probably a result of a deliberate decision taken at the corporate HQ to have a certain degree of formal decentralization, thus giving the subsidiaries more freedom to act on their own behalf. But the high degree of autonomy also reflects the high level of ignorance at HQ level. As noted earlier, ThermoSafe is a specialist business niche in SCA, and its market conditions unique among the SCA units. According to SCA America's top management, ThermoSafe has never been high on the priority list of the corporate HQ, since the company is doing well and SCA faces other problems. The perceived autonomy at the subsidiary level is therefore as much a result of ignorance and lack of interest at the corporate HQ level as it is a result of explicit formalization of responsibility and decision levels: *consequently, the autonomy is enjoyed in a situation of high uncertainty*. It is not clear what action ThermoSafe can take, because the autonomy is implicit rather than explicit. Or, put another way, the rules of the game for the subsidiary acting as an entrepreneur are vague, more or less in the same way as uncertainty among subordinates is much higher in a feudal system than in a bureaucracy (Perrow, 1986). The freedom to act based on a high degree of perceived autonomy is circumscribed by the high uncertainty linked to the unclear rules of the game. Autonomy as a dimension of entrepreneurial orientation is therefore a double-edged sword, depending on upon what the perceived autonomy is in fact based. This conclusion is especially relevant for the entrepreneurial orientation of a subsidiary, which by definition is part of a management system of an MNC organization. The link between autonomy and entrepreneurial orientation is therefore more complicated in the MNC/subsidiary case than for an independent firm, especially so if there is no explicit support from the corporate HQ, as in the case of ThermoSafe.

The last statement here leads us to consider risk-taking propensity as another important dimension of entrepreneurial orientation. We concluded above that ThermoSafe demonstrated a high degree of readiness to take on both business and financial risks. However, as part of, and dependent on, a larger organization, a subsidiary also encounters another type of risk, different from both business risk and financial risk. We may call this risk *organizational risk*. This risk is related to the legitimacy the subunit enjoys as part of the larger organization. The higher a subsidiary's legitimacy (as a participant of the MNC), the higher is the possibility of obtaining resources from the rest of the MNC, which are necessary for the survival and growth of the subsidiary. This legitimacy, in turn, is dependent on recognition of the subsidiary's competence by the corporate HQ and the other units. Although organizational legitimacy can be improved or damaged by subsidiaries' business, and in that sense can be related to the subsidiaries' business risk, it is above all a consequence of the fact that the subsidiary needs to foster its reputation in comparison with other subunits in the MNC.

In a situation where there is a high degree of autonomy in the MNC, it is obvious that the subsidiaries are highly dependent on market performance. Or, as expressed by Ken Harris: 'We are working our way through the market to get legitimate within the company – once we win the market we have to get credit from the rest of the organization.' It can very well be the case that a subsidiary finds a project too risky in terms of the possibility of losing too much credibility within the organization, even though business and financial risks are tolerable. Therefore, for a subsidiary in an MNC, entrepreneurial orientation is not only dependent on business and financial risks linked to the problems of assessing the future markets and the level of resource commitment, but also what impact the project will have on the subsidiary's legitimacy in the MNC; that is, its organizational risk.

The organizational risk is connected to the position of the subsidiary in relation to other subunits in the MNC. Another aspect of this relationship is the power structure within the MNC. In many ways, the MNC can be conceptualized as a federation, meaning that power is redistributed because no single person and no single group can be all-knowing and all-wise (Handy, 1993). That also means that influence 'tend[s] to be more contested within the MNC' (Ghoshal and Bartlett, 1990). As a consequence of this, every initiative by a subsidiary can be expected to be evaluated from the standpoint of its impact on other subunits' business. In the ThermoSafe case there are plenty of examples of that. One major obstacle to the subsidiary's entry into Europe is the 'wall blocking our overseas plans' created by subsidiary managers in Europe. The opposition from the European subsidiaries partly reflects the fact that ThermoSafe's plans differ from the local-for-local character of SCA's packaging business in Europe. An implementation of ThermoSafe's plans in Europe will threaten some of the subsidiaries' business there. It is also somewhat problematic that these changes originate from an 'outsider' in the SCA family, a newly created subsidiary in the USA with an unfamiliar business. Another key example of the resistance comes from HQ, which seems concerned about ThermoSafe 'going global', an issue that may related to the HQ's concern for 'power retention' (Yamin and Forsgren, 2006).

While autonomy and risk-taking as determinants for the entrepreneurial orientation of an MNC subsidiary need a somewhat different treatment compared to the case of an independent firm, the impact of the power game on the MNC would constitute a sixth dimension. It is clear that a possible counter-action by other subunits can have a negative impact on a subsidiary's entrepreneurial orientation. Even if a subsidiary enjoys a high degree of autonomy, is innovative, prepared to take risks, proactive and aggressive in relation to customers, a possible new entry can be delayed or abandoned because of its possible impact on the power distribution in the MNC.

Note

1. This study was conducted in the USA and Sweden during 2005. A total of twenty interviews were conducted, from one to several hours in length. Respondents were informed of the purpose of the research prior to the interview. All interviews were tape-recorded and most of them transcribed. A 'courtroom' procedure was used, in which questions concentrated on facts and events rather than on respondents' inter-pretations, especially of others' actions (Eisenhardt, 1989). More than one data source was used in the study: both quantitative and qualitative data from semi-structured interviews was collected. E-mails and telephone calls were made to follow up inter-views, and archival data – including company websites, business publications and other materials provided by the informants – were collected. Interviews were stopped when the authors considered that a level of understanding had been obtained that was satisfactory for the purpose of the research. On completion of all interviews, the gathered information was synthesized into a case history including, for example, a description of the decision-making process, actions that key managers took throughout the process, and the outcomes that followed.

References

Aldrich, Howard (1999) *Organizations Evolving* (London: Sage).

Andersson, Ulf and Mats Forsgren (1996) 'Subsidiary Embeddedness and Control in the Multinational Corporation', *International Business Review*, 5(5), 437–503.

Baird, I. S. and Thomas, H. (1985) 'Toward a Contingency Model of Strategic Risk Taking', *Academy of Science Review*, 10(2), 230–43.

Birkinshaw, Julian (1997) 'Entrepreneurship in Multinational Corporations: The Charac-teristics of Subsidiary Initiatives', *Strategic Management Journal*, 18(3), 207–29.

Birkinshaw, Julian and Jonas Ridderstråle (1999) 'Fighting the Corporate Immune System: A Process Study of Subsidiary Initiatives in Multinational Corporations', *International Business Review*, 8(2), 149–80.

Eisenhardt, K. M. (1989) 'Building Theories from Case Study Research', *Academy of Management Review*, 14(4), 532–50.

Ghoshal, Sumantra and Christopher Bartlett (1990) 'The Multinational Corporation as an Inter-organizational Network', *Academy of Management Review*, 15(4), 603–25.

Goodall, Keith and John Roberts (2003) 'Repairing Managerial Knowledge-Ability over Distance', *Organization Studies*, 24(7), 1153–75.

Handy, Charles (1993) 'Balancing Corporate Power: A New Federalist Paper', *Harvard Business Review*, 70(6), 59–72.

Lumpkin, G. T. and Gregory Dess (1996) 'Clarifying the Entrepreneurial Orientation Construct and Linking It to Performance', *Academy of Management Review*, 21(1), 135–72.

Perrow, Charles (1986) *Complex Organizations. A Critical Essay* (New York: Random House).

Shane, Scott (2000) *A General Theory of Entrepreneurship: The Individual-Opportunity Nexus* (Cheltenham: Edward Elgar).

Yamin, Mo (2002) Subsidiary Entrepreneurship and the Advantage of Multinationality, in V. Havila, M. Forsgren, and H. Håkansson (eds), *Critical Perspectives on Internationalization* (London: Pergamon Press).

Yamin, Mo and Mats Forsgren (2006) 'Hymer's Analysis of the Multinational Corporation: Power Retention and the Demise of the Federative MNE', *International Business Review*, 15(2), 166–79.

3
The Role of Sales Subsidiaries in MNC Innovativeness

Jani Lindqvist, Kirsimarja Blomqvist and Sami Saarenketo

Introduction

In this study we are interested in the role of sales subsidiaries on MNC innovativeness and related knowledge transfer. In knowledge-based competition, MNC competitiveness emerges from its ability to recognize the potential innovativeness of an idea or a method, and capitalize on this in worldwide markets through suitable transfer mechanisms. However, transferring the knowledge on ideas and innovations from locally embedded sales subsidiaries is challenging and a potential cause of anxiety in international business (see also Birkinshaw *et al.*, 2000).

Sales subsidiaries are a potential source of innovative ideas on ways of increasing customer loyalty and satisfaction, and for aggregated customer and market knowledge for innovations. Thus the relationship with sales subsidiaries plays a major part when a firm becomes 'market orientated', based on the corporate *capability to identify and transfer knowledge of customer needs* (Kohli and Jaworski, 1990) to enhance corporate innovativeness and performance (Erdil *et al.*, 2004; Hurley and Hult, 1998; Narver *et al.* 2000). In addition to new customer-driven product and service innovations, sales subsidiaries are sources of 'best practice' for customer interaction, which have the potential for administrative innovations providing additional means for MNCs to differentiate from competitors. Nevertheless, sales subsidiaries have been almost completely ignored as a part of the innovation apparatus of the MNC, and in the literature on innovations, where the focus has still been mainly on technical innovations rather than administrative and service innovations (see, for example, Damanpour, 1991). However, as service and administrative innovations are becoming increasingly important in customer-driven competition based on knowledge and relationships, the role of MNC sales subsidiaries has a specific, but as yet neglected, role.

Our case-study company Alpha is a MNC, and a global leader with its key products. It has long had a reputation for innovative products and services. It is operating in commodity markets, and facing an increasingly strong global competition. In addition to product and process innovations, Alpha is looking for new ways to leverage for competitiveness from continuous innovation. Innovation is highly important in the corporate strategy and the sales subsidiaries are expected to move from their traditional role in MNCs – that is, selling as much as they can – to creating relationships with their most profitable customers to gather knowledge for improved corporate innovation.

Our theoretical framework is built deductively from the knowledge-based view of the firm, and we focus on the challenges and possibilities of knowledge transfer of innovative ideas in MNC–subsidiary relations. Based on a literature review, we build a conceptual model, which will be discussed and analysed in relation to the inductive empirical study of knowledge transfer of four sales subsidiaries and the HQ of MNC Alpha. Finally, we discuss the results and conclude the chapter with some managerial implications and proposals for further research.

Literature review

According to the knowledge-based view (KBV) of the firm, a multinational corporation (MNC) can be understood as a social community specializing in speed and efficiency in the creation and transfer of knowledge across country borders, where it excels when compared to markets (Almeida *et al.*, 2002; Kogut and Zander, 1996). We understand knowledge as being composed of know-how and information, where know-how is 'the accumulated practical skill or expertise that allows one to do something smoothly and efficiently' (von Hippel, 1988), and information is 'knowledge which can be transmitted without loss of integrity once the syntactical rules required for deciphering it are known' (Kogut and Zander, 1992). The nature of knowledge can be tacit or explicit. *Explicit knowledge* can be expressed and codified. *Tacit knowledge* is personal, context-dependent and based on practice and experience. It has been argued that tacit knowledge can only be transmitted to others by sharing mutual experiences and active participation in real-time, face-to-face interaction (Nonaka and Takeuchi, 1995; see also Pöyhönen and Blomqvist, 2006).

MNC typically requires the application of many types of knowledge, and the extent of knowledge is usually not specific to one application or organizational unit, but for efficient use, knowledge 'exports' and 'imports' within MNC are needed. Traditionally, however, MNCs have created knowledge in the HQ and then transferred it to the subsidiaries in the form of processes or products (Almeida *et al.*, 2002). Bartlett and Ghoshal (2002) suggest a transnational view, in that a worldwide company should utilise all the knowledge from various parts of the organization to innovate through organizational learning. However, MNC is rarely a homogeneous social community with common organizational knowledge transfer, but rather a heterogeneous collection of worldwide scattered identities (Forsgren, 2006).

The local knowledge that subsidiaries absorb in the markets can both enhance the competitive advantage and performance of the subsidiary in its own market, and enhance MNC competitiveness through transfer of local knowledge to other corporate units (Andersson *et al.*, 2001; Zander and Kogut, 1995). Local knowledge created in subsidiaries can thus be very valuable, if it is rare and not imitable – for example, tacit and contextual local knowledge. Knowledge is produced and reproduced in a social setting, and is not fully reducible to individuals (Foss, 1996). Such *collective knowledge* is seen strategically as the most significant kind of organizational knowledge (Conner and Prahalad, 1996; Tsoukas, 2002). This is in line with the KBV approach to innovations, which are seen to emerge through a social process with interactive relations and complex feedback mechanisms, and facilitated by the individual and collective expertise of the company's employees (Kline and Rosenberg, 1986; Leonard and Sensiper, 1998). For example, product and service decisions regarding design or style require sensitivity to diverse norms and attitudes in the global setting.

To summarize, MNCs need to manage knowledge on a global level (Andersson *et al.*, 2001; Subramaniam and Venkatraman, 2001). Despite the importance of the topic for multinational corporations, there are only few studies available in the literature that focus in the transfer of market knowledge in MNCs (Bennett and Gabriel, 1999; Schlegelmilch and Chini, 2003; Simonin, 1999). Our study contributes in particular to the knowledge-based view of the MNC by focusing on the less-studied role of sales subsidiaries, and the knowledge transfer from subsidiary to HQ.

A conceptual model of knowledge transfer between MNC sales subsidiary and HQ

According to Narver *et al.* (2000), *market orientation* is positively correlated with innovativeness, business profitability, sales growth and new-product success. To be able to reap the benefits of being market-orientated, it has been suggested that a firm needs (i) to generate market information about the expressed and latent needs of customers and external environmental factors; (ii) transfer such information between organizational functions; and (iii) eventually develop and implement strategies in response to the acquired information (Kohli and Jaworski, 1990).

A simplified version of the role of sales subsidiaries as potential sources of market-driven innovation in the MNC is illustrated in Figure 3.1. The figure

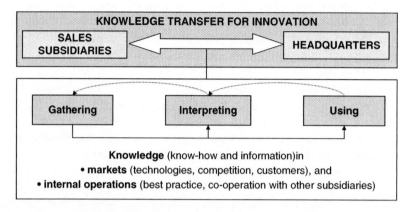

Sources: Day, 1994; Kogut and Zander, 1993.
Figure 3.1 Sales subsidiaries as sources of market-driven innovation

highlights the activities necessary to enable a MNC to benefit from available knowledge: gathering knowledge, and interpreting and using it. The knowledge transfer is portrayed as bi-directional, because responsibility for the process is often shared, depending on the purpose and perspective, as well as the subject and content of the information processing (see Egelhoff, 1991).

In the customer interface, the ability of a firm to create and manage close customer relationships is increasingly important in accessing available market information (on both customers and competition), as it may lead to unique market opportunities and ultimately to innovation (Nonaka and Takeuchi, 1995). Day (1994) refers to such activities as *market sensing* and *customer-linking* capabilities.

In the following, we review the literature on the specific issues and challenges in MNC knowledge transfer in the sales subsidiary–HQ relationship.

Source and recipient of knowledge

Almeida *et al*. (2002) emphasize that all MNC units, simultaneously and inter-actively, engage in the creation, acquisition (learning by doing, external learning) and application of knowledge. Therefore, knowledge transfer should be approached from the dual perspectives of recipient and source of knowledge.

For effective knowledge transfer, the knowledge source needs to understand the knowledge requirements of the recipient. In a similar vein, it is important for the source to be able to identify the appropriate knowledge to be transferred (Birkinshaw *et al*., 2000). Relevant knowledge may relate to customers and other external relationships, internal production or research and development (R&D), or to knowledge of local clusters – for example, research institutions and well-educated workforce. It is assumed that externally gained knowledge is more difficult to transfer than internally developed knowledge because of the recipient's fewer direct contacts (Foss and Pedersen, 2002).

In order to assimilate new knowledge, prior related knowledge is needed. This knowledge determines a firm's *absorptive capacity* – that is, its 'ability to recognise the value of new, external information, assimilate it, and apply it to commercial ends' (Cohen and Levinthal, 1990). In fact, absorptive capacity has been proposed as the most significant determinant of MNC internal knowledge transfer (Gupta and Govindarajan, 2000). Also, Szulanski (1996) claims that inability to identify and to transfer best practice within the firm is primarily because of the recipient's lack of absorptive capacity, causal ambiguity, and arduous relationships. *Retentive capacity* depicts the recipient's ability to implement received knowledge, while *unprovenness* of knowledge may hinder its transfer and the recipient's motivation to absorb and implement it.

Perception gaps are differences in interests and perception between the parties, such as between the HQ and subsidiaries (see Asakawa, 2001; Birkinshaw *et al.*, 2000). Subsidiaries may over- or underestimate the strategic importance of their roles in relation to HQ management perceptions. Birkinshaw *et al.* (2000) propose that high perception gaps may lead to a vicious circle where a high degree of control leads to low co-operation, and subsequently to higher perception gaps. Perceptual congruence enhances communication and feelings of inter-personal satisfaction.

Motivation and attitude – for example, *reluctance/NIH* (not invented here) – also influence knowledge transfer. Lack of motivation can hamper both knowledge transfer and its implementation. Benefits that parties experience are important for effective knowledge transfer (Gupta and Govindarajan, 2000; Schlegelmilch and Chini, 2003). Both the recipient and the source of knowledge should experience knowledge transfer beneficially.

The nature and characteristics of knowledge

The nature and characteristics of knowledge have an impact on knowledge absorption, creation, transfer and implementation. Tacit, contextual and experiential knowledge is more valuable than explicit knowledge, as it is difficult for competitors to imitate. On the other hand, it is also difficult to recognize, make explicit and transfer such knowledge within the company.

In addition to having easier access to local market information – for example, through local contacts or staff with specific language skills – the knowledge of the sales subsidiaries emerges from a specific cultural context and is experiential in nature, thus making the knowledge 'sticky' – that is, difficult and costly to transfer (Nonaka and Takeuchi, 1995; Schlegelmilch and Chini, 2003; Szulanski, 1996). Salespeople may also gain knowledge of research spillovers (see Ostry and Gestrin, 1993), which have a significant potential for enhancing MNC innovativeness.

Causal ambiguity refers to the inability to determine the reasons for success or failure in knowledge creation, even *ex post*. It may result from poorly understood yet distinctive features of the new context where the knowledge has been applied, or from the fact that much of the knowledge is embodied in tacit and collectively held human skills (Szulanski, 1996).

Resources are needed for efficient knowledge transfer. Resource costs for transferring knowledge may consist of time, money or suitable tools and processes for knowledge transfer (Foss and Pedersen, 2002). Availability of slack resources in a subsidiary has been purported to promote innovativeness in terms of creating and diffusing innovations (Ghoshal and Bartlett, 1988).

Global context

One of the key challenges to knowledge transfer in a MNC emerges from 'the problem of organisation in a setting of physical separation through time and space and separation of key members by culture and language' (Schlegelmilch and Chini, 2003). Organizational distance increases both anxiety and ambiguity – that is, the understanding of the logical linkages between actions and outcomes, inputs and outputs. It can be proposed that the less organizational distance there is between units, the greater is the effectiveness of knowledge transfer.

Knowledge transfer process

Feedback has been identified as a critical element in MNC knowledge creation, transfer and innovativeness (Kline and Rosenberg, 1986). In the sales subsidiary–HQ relationship they both act as sender–receivers, and we consider that feedback loops are important in the exchange of tacit knowledge typical in a marketing and sales context. In similar vein, special infrastructure is needed to support knowledge transfer. For tacit knowledge, wider communication bandwidth and non-routine communication, such as cross-unit temporary teams, communities of practice and job rotation may be needed for its successful transfer (see, for example, Foss and Pedersen, 2002).

Control and co-ordination mechanisms

Control, co-ordination and co-operation are important elements in HQ-subsidiary relationships and essential for MNC knowledge transfer and innovation (Ghoshal and Bartlett, 1988). The means of exercising control are monitoring, directing, evaluating and rewarding (Anderson and Oliver, 1987). Subsidiary characteristics, task complexity variables and the nature and locus of knowledge affect various control combinations. The more locally embedded the tacit knowledge, the higher the need for subsidiary autonomy. On the other hand, more successfully the MNC's internally-accumulated knowledge may be transferred, the higher the interdependence between the transferring and receiving units (Foss and Pedersen, 2002).

Standardization and bureaucratic routines as co-ordination mechanisms have been suggested to inhibit communication and innovation (Narver *etal.*, 2000). Strong and active control may be considered to be sign of lack of trust, which hampers co-operation, and through that also innovation, as trust and co-operation are vital for both knowledge creation and innovation (Birkinshaw *etal.*, 2000). Behavioural-based control, it has been suggested, is positively related to organizational commitment (Baldauf *etal.*, 2005) and to innovative and more supportive organizational cultures (Oliver and Anderson, 1994). Trust and shared vision are examples of informal and behavioural-based control. Standardization and trust have both been suggested as important determinants of effective knowledge transfer (Li, 2005), and can be seen as alternative or complementary governance mechanisms for control, subsequently affecting knowledge transfer between subsidiary and HQ. Trust has also been identified as a critical principle through which organizations arrange and co-ordinate their activities (McEvily *etal.*, 2003). Shared vision has been identified as promoting the interests of the MNC as a whole rather than the individual units (see, for example, Tsai and Ghoshal, 1998).

In Figure 3.2 we synthesize the literature review into a conceptual model describing the key elements in the specific issues and challenges in knowledge transfer between sales subsidiaries and HQ. Next, the conceptual model based on earlier analytical and empirical research is illustrated with empirical data gathered inductively from case-study firm, Alpha.

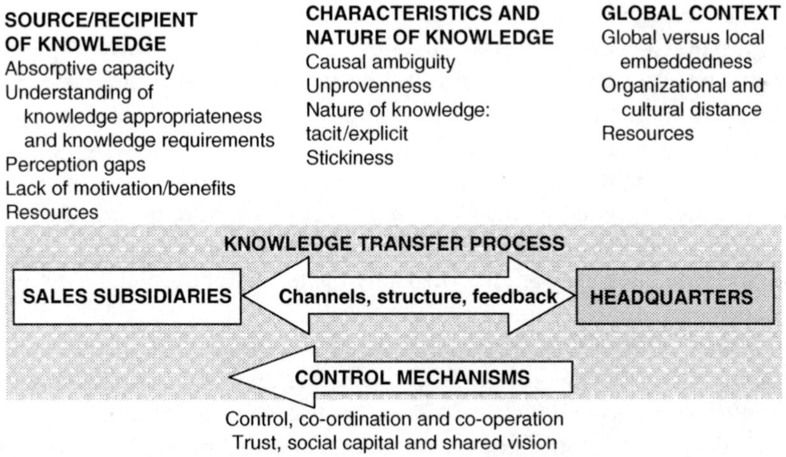

SOURCE/RECIPIENT OF KNOWLEDGE
Absorptive capacity
Understanding of
 knowledge appropriateness
 and knowledge requirements
Perception gaps
Lack of motivation/benefits
Resources

CHARACTERISTICS AND NATURE OF KNOWLEDGE
Causal ambiguity
Unprovenness
Nature of knowledge:
tacit/explicit
Stickiness

GLOBAL CONTEXT
Global versus local
 embeddedness
Organizational and
 cultural distance
Resources

KNOWLEDGE TRANSFER PROCESS

SALES SUBSIDIARIES ← Channels, structure, feedback → **HEADQUARTERS**

CONTROL MECHANISMS

Control, co-ordination and co-operation
Trust, social capital and shared vision

Figure 3.2 A conceptual model of the specific issues and challenges in knowledge transfer between sales subsidiaries and HQ

Research design, data collection and analysis

In order to increase understanding of the factors affecting knowledge transfer between a sales subsidiary and headquarters, a qualitative methodological approach was undertaken. First, we describe our data collection process, after which we depict the activities that relate to knowledge creation and transfer in the sales subsidiary–HQ relationship. Finally, we compare our empirical results with the theory-based model developed above.

Data collection

This study is based on a qualitative database collected in nine focus group (FG) interviews, conducted in four sales subsidiaries (two each) and the HQ of a MNC, Alpha. FG is a special type of group interview that has a predefined purpose, size, composition and procedure. The focus group method permits the putting of open-ended questions, which allow the experience of participants to emerge (Hines, 2000). In the nine distinct focus group interviews there were forty-nine participants in total, of which six were people from the HQ and the remaining forty three from the subsidiaries. The purpose was to form the groups so that they included a relatively homogenous selection of participants representing the same managerial level, but from different divisions of the company.

Three interviews were carried out in subsidiaries in Western and Southern Europe, with one was carried out in the USA; the HQ interview was done in Northern Europe. Each FG was conducted on the company property, with four to eight participants plus the facilitator from the research team.

All FGs followed the same structure: first, for orientation and warm-up, the study was introduced, and the participants introduced themselves, describing their perspective on the innovativeness of Alpha. The theme questions concerned: factors affecting innovativeness; how is innovativeness visible; how the circumstances for innovativeness could be improved; an account of innovations that emerged or would be emerging from the sales subsidiary; plus specific questions regarding the knowledge transfer between the sales subsidiary and the HQ. Finally, there was an open-floor meeting for comments and discussion. Participants were first allowed a few minutes to think the each issue, and write down key points. Thereafter the facilitator opened the discussion for all on each specific theme.

The FG session theme questions were used as research questions during the primary coding. In the second phase of the analysis, researchers defined the relationship between primary codes. This was done by reading through the primary data and exploring the relationships expressed by the FG participants. In the second phase, researchers defined substantial codes to which the primary codes are related. Next, we shall present the factors affecting knowledge transfer according to the substantial codes.

Data analysis

We first carry out some more general analysis and then reflect the data to the conceptual model.

Nature of knowledge

Salespeople discovered that there was a plenty of explicit knowledge requirements from the HQ in the form of sales documents, filled in routinely into corporate IT-systems. Moreover, there was frustration concerning the sharing of explicit knowledge between sales subsidiaries; and, added to this the staff of the sales subsidiaries were not satisfied with the amount and quality of feedback from HQ. The communication was not experienced as mutual and satisfactory.

Tacit knowledge seemed to have been transferred in lesser amounts, even though knowledge on changes in customer preferences and culture is mainly tacit. In particular, informal events and meetings were considered beneficial for intra-firm and customer-related tacit knowledge transfer. Even so, there were challenges in sharing the tacit knowledge with remote units – that is, other sales subsidiaries and the HQ.

Knowledge transfer between sales subsidiary and customer

As we enquired about the different aspects of knowledge transfer process to and from sales subsidiaries for innovativeness, it became evident that the *types of information* gathered from customers is diverse. The knowledge that was gathered and transferred concerned corporate day-to-day business operations consisting of knowledge regarding customer satisfaction, quality and schedules. There were few comments about knowledge regarding the events and trends in local markets in general. The lack of more value-adding knowledge was mainly attributed to excess supply in the market and tough price competition.

Knowledge-access methods were customer visits and telephone calls. In addition, customer satisfaction surveys were seen as an important part of knowledge retrieval. The salespeople aimed at getting several contact points among their customers, as this was considered to be a way of promoting sales of more value-adding services. The knowledge gathered during the interaction often relates to complaints, pricing issues and service satisfaction. The primary uses for this information are for the improvement of processes between the customer and the firm, planning for future business with the customer, following competitor activities, and the improvement of services.

In the knowledge transfer between customers and salespeople there is also a pull factor for innovation: customers often contact the sales subsidiary when they have problems with Alpha's products or services, or are making changes to their own processes that will affect the common interface. Salespeople value highly the knowledge they gain from this interaction, as it is unique and difficult to access by other means. However, the sales subsidiaries consider reaction to such signals as being hampered by the slow feedback from the HQ.

Knowledge transfer between sales subsidiary and HQ

In Alpha's internal communications knowledge transfer with the HQ often remains at a minimum, because it is felt that there is little value in the relationship. The global perspective of HQ adds to this threshold, because the sales subsidiaries feel that they do not get full support from HQ. In addition, there are some problems with interpreting all the knowledge coming from HQ. Based on focus groups at HQ, several contact points are available for good communication. Yet there is lack of follow-up regarding how the knowledge is disseminated and utilized, in both the sales subsidiary and HQ. Also, there is no bi-directional and mutual knowledge transfer, as the information flows are primarily from the HQ to sales subsidiary and very little takes place in the other direction (apart from explicit knowledge transfer through IT systems). Finally, there is an abundance of information transmitted to the sales subsidiaries, which diminishes the perception of its relevance among the salespeople.

Knowledge transfer between sales subsidiary and other MNC units

Salespeople discussed (also unprompted) that in their internal networks the factories are very co-operative and responsive to the sales subsidiary's needs despite the typically long physical and psychical distance between them. Instead, when communicating between distributed functions there are a few problems, in particular concerning global account management. Sales subsidiaries perceive that global accounts are treated too much in general terms, and by too many people, without the needed customization of services. As a result, there is much inefficiency in knowledge transfer regarding global accounts.

Types of innovation

Innovations by Alpha that the sales subsidiaries consider themselves to have facilitated were primarily service innovations. There were also many administrative innovations. However, it became apparent in the interviews that most incremental innovations are overlooked, and knowledge of them is rarely exchanged with other than the closest colleagues. The impact of the sales subsidiaries' knowledge on the creation of innovations in production processes and products was reported to be quite low.

Next we shall reflect back to the data from the focus group interviews with four sales subsidiaries and the HQ of MNC Alpha, to illustrate the important but often neglected role of sales subsidiaries in the innovation apparatus of a MNC.

Sources and recipients of knowledge

Based on our focus groups, we assume that there is *no clear understanding of knowledge needs* for MNC innovativeness in sales subsidiaries. We also recognize that the knowledge sales subsidiaries perceived as being *appropriate and valuable* for innovation was limited, and very little knowledge about ideas for improving business together was transferred.

Absorptive capacity, perception gaps and motivation seemed to be closely linked. Despite corporate emphasis on innovation, the *sales subsidiaries were not able to recognize the value of the customer-related knowledge, as innovation was not emphasized in their roles*. Therefore the absorptive capacity within the sales subsidiaries may be limited because of a lesser emphasis on their role in corporate innovation. This may also be related to their *motivation*, and unwillingness to transfer knowledge, as, in their experience, they did not receive sufficient recognition from HQ. It is also evident that salespeople did not gain enough *personal benefits* from participating in the knowledge transfer for innovation. Plausibly, retaining knowledge was considered to be a source of power, and giving it up was therefore not an attractive option.

It appears that knowledge transfer and retention were strong at the level of specific individuals, but not at company level. Sales subsidiaries and HQ experienced the stage and level of knowledge transfer as occasional and dependent on personal relationships and informal meetings, rather then as part of a managed process.

Characteristics and the nature of knowledge

Tacit and experiential knowledge was found to be sticky, demanding interpersonal and face-to-face meetings to allow it to be transferred. The slow transfer of *explicit knowledge* makes for an additional hindrance: whether the requested responses relate to operative developments or new ideas, they come after a considerable lag. Improvements in communication across the whole corporation was desirable.

Unprovenness of knowledge was noticeable as reluctance/NIH and trust were mentioned, mainly in the context that HQ was perceived as being interested only in the largest customers, or that there was no feedback for ideas generated among sales subsidiaries. Unprovenness of knowledge may also have hindered knowledge transfer and motivation of the recipients to absorb and implement the knowledge, as some respondents mentioned the lack of trust between salespeople and HQ, especially regarding communication about market knowledge. In many discussions, tacit and individually held knowledge was found to be very useful and highly appreciated, whereas excess amounts of explicit knowledge were transferred from HQ to the sales subsidiaries.

Global context

Physical separation across time and space, as well as the separation of key members because of culture and language, were also highlighted in focus groups (see Schlegelmilch and Chini, 2003). This potential strength of a MNC also cause anxiety. Separation of the US sales subsidiary and European units was most evident. None the less, salespeople in the different European sales subsidiaries also experienced difficulties in communicating with each other, which they contributed to both the time difference and cultural dissimilarities. Sales subsidiary staff also lacked *resources*, perceiving that they did not have sufficient time for gathering, interpreting, transferring and using knowledge for innovation.

The knowledge-transfer process

The transfer of explicit knowledge was not considered to be efficient, mainly because of a lack of feedback from the HQ. It seems that knowledge transfer for innovation in the sales subsidiary–HQ relationship is still very much in its initial stages. A *communication gap* existed between the source and the recipient, as well as the adaptation of practices to the recipient's needs. It also seems that participants lacked suitable *channels* for tacit knowledge transfer. *Infrastructure* available was modest, and there was a demand for champions to enhance tacit knowledge transfer.

Control and co-ordination mechanisms

In Alpha, there was co-ordination and control, but somewhat less co-operation in HQ–subsidiary relationships, and there were few suitable means for rewarding innovation among sales subsidiaries. The subsidiaries' performance was monitored but, based on the focus groups, there was not enough performance orientation, and under-delivery was not penalized. The HQ's means of control consisted of shared processes, information and rules. Because much of the *information was uni-directional*, this may have caused some decrease in trust in HQ's the goodwill and active support for sales subsidiaries. In general, relationships with HQ were not always very trusting and close, but have been described as 'arduous'. Closer relationships, with bi-directional communication and trust, could have enhanced knowledge transfer and helped the subsidiaries to identify themselves with some of the market-driven product innovations Alpha had introduced. There may also have been a lack of shared vision, as there were some feelings of '*us against them*' among the sales subsidiaries. Sometimes relationships with HQ were perceived as being costly and non-productive, resulting in a vicious circle of less communication and consequently less knowledge transfer.

Summary and discussion

From the perspective of innovation and MNC, knowledge transfer of novel ideas, best practice and innovative solutions have emerged among the primary sources of competitive success for international corporations (see, for example, Bartlett and Ghoshal, 2002). The creation, identification and transfer, not to mention the utilization, of such knowledge imposes challenges for which multinational firms are rarely ready. In this chapter we have studied the role that a sales subsidiary is capable of playing in the innovation apparatus of a MNC, and developed a model of knowledge transfer between subsidiaries and their HQ. We have highlighted issues affecting the transfer of knowledge gathered by foreign sales subsidiaries that could be harnessed for improved corporate market orientation and innovation.

Control and co-ordination mechanisms were given a surprising large amount of attention in the FG interviews. On the one hand, behavioural control appears to be desirable for increasing tacit knowledge transfer, but on the other hand, the global co-ordination of activities was experienced as decreasing the initiative and creativity of subsidiaries. Clearly, there were some *perception gaps* in the sales subsidiary role in MNC innovation. Supposedly, subsidiaries have under-estimated their *strategic role* in corporate innovation and overestimated their role as independent decision-makers desiring more feedback from HQ. If so, it seems to be the beginning of a vicious circle of decreasing commitment in customer-driven innovation.

Apparently, there was too little willingness to share knowledge, and it was found to be difficult to make appropriate contacts with whom to share knowledge on innovative ideas, and to develop them further. In particular, the difficulties experienced in the transfer of knowledge between sales subsidiaries, and between individual sales subsidiaries and HQ suggest that Gupta and Govindarajan (2000) were correct in stating that there are fewer frictions when complementary knowledge is transferred (sales subsidiary–factory) than in the context of substitutive knowledge transfer (sales subsidiary–sales subsidiary). It is proposed that explicit and MNC internally-accumulated knowledge may be transferred more successfully, the higher the interdependence between the transferring and receiving units (see also Foss and Pedersen, 2002). Based on the FGs, it seems that in Alpha's situation the most important method for knowledge transfer for innovation is the possibility of communicating regularly and informally with colleagues. This finding also corresponds with previous studies, where it has been noted that internal communication increases diversity, and facilitates a better flow and dissemination of ideas inside the organization (Aiken and Hage, 1971). Unfortunately, there were not so many possibilities for informal knowledge transfer, because of resource constraints and lack of support structures.

Conclusions

MNC innovativeness and knowledge advantage emerges from the possibility of transferring tacit knowledge between an organization's units that is not similarly feasible in inter-organizational relationships and related governance. In our case-study firm, Alpha, this may have been affected by cultural distance and perception gaps between subsidiaries and the HQ, which could be alleviated by appropriate incentives and communication (see also Birkinshaw *et al.*, 2000). It is further proposed that the efficient transfer of explicit knowledge requires both bi-directional knowledge transfer structures and global integration.

Based on our study, it seems that, in addition to the literature on market orientation, social capital theory (Nahapiet and Ghoshal, 1998) could provide a promising research approach to knowledge transfer issues between a MNC and a sales subsidiary. The more complex the knowledge to be integrated, the more systematic, refined and versatile are the mechanisms that are needed. We therefore, suggest that, at the operative level, the managerial task is to create conditions where individuals can exchange their idiosyncratic knowledge, while at the corporate level this must be supported through suitable formal structures and informal culture – for example, introducing champions for knowledge transfer.

Moreover, we have identified several challenges in MNC – subsidiary knowledge transfer through the focus group interviews. Thus focus group discussion appears to be a powerful method for unearthing sources of stakeholder anxiety in a worldwide operating corporation.

Finally, we feel that the present study, in spite of its limitations, offers an important extension to our current knowledge on the role of sales subsidiaries in MNC innovativeness. A thorough understanding of the innovation apparatus of an MNC continues to be of great importance to academics and practitioners in the field of international business.

References

Aiken, Michael and Jerald Hage (1971) 'The Organic Organization and Innovation', *Sociology*, 5(1), 63–82.

Almeida, Paul, Jaeyong Song and Robert M. Grant (2002) 'Are Firms Superior to Alliances and Markets?', *Organization Science*, 13(2), 147–61.

Anderson, Erin and Richard L. Oliver (1987) 'Perspectives on Behavior-based versus Outcome-based Salesforce Control Systems', *Journal of Marketing*, 51(4), 76.

Andersson, Ulf and Mats Forsgren (2000) 'In Search of Centre of Excellence: Network Embeddedness and Subsidiary Roles in Multinational Corporations', *Management International Review*, 40(4), 329–50.

Andersson, Ulf, Ingmar Björkman and Mats Forsgren (2005) 'Managing Subsidiary Knowledge Creation: The Effect of Control Mechanisms on Subsidiary Local Embeddedness', *International Business Review*, 14(5), 521–38.

Andersson, Ulf, Mats Forsgren and Ulf Holm (2001) 'Subsidiary Embeddedness and Competence Development in MNCs – a Multi-level Analysis', *Organization Studies*, 22(6), 1013–34.

Andersson, Ulf, Mats Forsgren and Ulf Holm (2002) 'The Strategic Impact of External Networks: Subsidiary Performance and Competence Development in the Multinational Corporation', *Strategic Management Journal*, 23(11), 979–96.

Asakawa, Kazuhiro (2001) 'Organizational Tension in International R&D Management: The Case of Japanese Firms', *Research Policy*, 30(3), 735–57.

Baldauf, Artur, David W. Cravens and Nigel F. Piercy (2005) 'Sales Management Control Research – Synthesis and an Agenda for Future Research', *Journal of Personal Selling & Sales Management*, 25(1), 7–26.

Bartlett, Christopher A. and Sumantra Ghoshal (2002) *Managing Across Borders – The Transnational Solution*, 2nd edn (Boston, Mass.: Harvard Business School Press).

Bennett, Roger and Helen Gabriel (1999) 'Organisational Factors and Knowledge Management within Large Marketing Departments: An Empirical Study', *Journal of Knowledge Management*, 3(3), 212–25.

Birkinshaw, Julian (2002) 'Managing Internal R&D Networks in Global Firms – What Sort of Knowledge is Involved?', *Long Range Planning*, 35(3), 245–67.

Birkinshaw, Julian, Ulf Holm, Peter Thilenius and Niklas Arvidsson (2000) 'Consequences of Perception Gaps in the HQ – subsidiary relationship', *International Business Review*, 9(3), 321–44.

Cantwell, John and Ram Mudambi (2005) 'MNE Competence-creating Subsidiary Mandates', *Strategic Management Journal*, 26(12), 1109–28.

Cohen, Wesley M. and Daniel A. Levinthal (1990) 'Absorptive Capacity: A New Perspective on Learning and Innovation', *Administrative Science Quarterly*, 35(1), 128–52.

Conner, Kathleen R. and C. K. Prahalad (1996) 'A Resource-based Theory of the Firm: Knowledge versus Opportunism', *Organization Science*, 7(5), 477–501.

Cravens, David W. (1995) 'The Changing Role of the Sales Force', *Marketing Management*, 4(2), 49–57.

Damanpour, Fariborz (1991) 'Organizational Innovation: A Meta-analysis of Effects of Determinants and Moderators', *Academy of Management Journal*, 34(3), 555–90.

Day, George. S. (1994) 'The Capabilities of Market-driven Organizations', *Journal of Marketing*, 58(4), 37–52.

Egelhoff, William G. (1991) 'Information-processing Theory and the Multinational Enterprise', *Journal of International Business Studies*, 22(3), 341–68.

Erdil, Sabri, Oya Erdil and Halit Keskin (2004) 'The Relationships between Market Orientation, Firm Innovativeness, and Innovation Performance', *Journal of Global Business and Technology*, 1(1), 11 pp.

Fahey, Liam and Laurence Prusak (1998) 'The Eleven Deadliest Sins of Knowledge Management', *California Management Review*, 40(3), 265–76.

Forsgren, Mats (2006) 'Are Multinationals Superior or Just Powerful? A Critical Review of the Evolutionary Theory of the MNC', Keynote presentation given at 33rd AIB-UK Conference, 7 April 2006, Manchester.

Foss, Nicolai J. (1996) 'Knowledge-based approaches to the Theory of the Firm: Some Critical Comments', *Organization Science*, 7(5), 470–6.

Foss, Nicolai J. and Torben Pedersen (2002) 'Transferring Knowledge in MNCs: The Role of Sources of Subsidiary Knowledge and Organizational Context', *Journal of International Management*, 8(1), 49–67.

Ghoshal, Sumantra and Christopher A. Bartlett (1988) 'Creation, Adoption and Diffusion of Innovations by Subsidiaries of Multinational Corporations', *Journal of International Business Studies*, 19(3), 365–88.

Ghoshal, Sumantra and Lynda Gratton (2002) 'Integrating the Enterprise', *MIT Sloan Management Review*, 44(1), 31–8.

Grant, R (1996) 'Toward a Knowledge-based View of the Firm', *Strategic Management Journal*, 17 (Winter special issue), 109–22.

Gupta, Anil K. and Vijay Govindarajan (2000) 'Knowledge Flows within Multinational Corporations', *Strategic Management Journal*, 21(4), 473–96.

Hines, Tony (2000) 'An Evaluation of Two Qualitative Methods (Focus Group Interviews and Cognitive Maps) for Conducting Research into Entrepreneurial Decision Making', *Qualitative Market Research: An International Journal*, 3(1), 7–16.

Hurley, Robert F., G. Thomas, M. Hult (1998) 'Innovation, Market Orientation and Organizational Learning: An Integration and Empirical Examination', *Journal of Marketing*, 62(3), 42–54.

Kline, Stephen and Nathan Rosenberg (1986) 'An Overview on Innovation', in Ralph Landau and Nathan Rosenberg (eds), *The Positive Sum Strategy – Harnessing Technology for Economic Growth* (Washington, DC: National Academic Press).

Kogut, Bruce and Udo Zander (1992) 'Knowledge of the Firm, Combinative Capabilities, and the Replication of Technology', *Organization Science*, 3(3), 383–97.

Kogut, Bruce and Udo Zander (1993) 'Knowledge of the Firm and the Evolutionary Theory of Multinational Corporations', *Journal of International Business Studies*, 24(4), 625–45.

Kogut, Bruce and Udo Zander (1996) 'What Firms Do? Coordination, Identity and Learning', *Organization Science*, 7(5), 502–18.

Kohli, Ajay K. and Bernard J. Jaworski (1990) 'Market Orientation: The Construct, Research Propositions, and Management Implications', *Journal of Marketing*, 54(2), 1–18.

Leonard, Dorothy and Sylvia Sensiper (1998) 'The Role of Tacit Knowledge in Group Innovation', *California Management Review*, 40(3), 112–32.

Li, Li (2005) 'The Effects of Trust and Shared Vision on Inward Knowledge Transfer in Subsidiaries' Intra- and Inter-organizational Relationships', *International Business Review*, 14(1), 77–95.

McEvily, Bill, Vincenzo Perrone and Akbar Zaheer (2003) 'Trust as an Organizing Principle', *Organization Science*, 14(1), 91–103.

Narver, John C., Stanley F. Slater and Douglas L. MacLachlan (2000) 'Total Market Orientation, Business Performance, and Innovation', Working paper, Marketing Science Institute Working Paper Series, Report No. 00–116.

Nonaka, Ikujiro and Hirotaka Takeuchi (1995) *The Knowledge Creating Company* (New York: Oxford University Press).

Oliver, Richard L. and Erin Anderson (1994) 'An Empirical Test of the Consequences of Behavior- and Outcome-based Sales Control Systems', *Journal of Marketing*, 58(4), 53–67.

Ostry, Sylvia and Michael Gestrin (1993) 'Foreign Direct Investment, Technology Transfer and the Innovation-Network Model', *Transnational Corporations*, 2(3), 7–30.

Pöyhönen, A. and K. Blomqvist (2006) 'Knowledge-based View of the Firm – Foundations, Focal Concepts and Emerging Research Issues', Paper to be presented at 7th ECKM Conference on Knowledge Management, Budapest, 2006.

Schlegelmilch, Bodo B. and Tina Claudia Chini (2003) 'Knowledge Transfer Between Marketing Functions in Multinational Companies: A Conceptual Model', *International Business Review*, 12(2), 215–32.

Simonin, Bernard L. (1999) 'Transfer of Marketing Know-how in International Strategic Alliances: An Empirical Investigation of the Role and Antecedents of Knowledge Ambiguity', *Journal of International Business Studies*, 30(3), 463–90.

Subramaniam, Mohan and N. Venkatraman (2001) 'Determinants of Transnational New Product Development Capability: Testing the Influence of Transferring and Deploying Tacit Overseas Knowledge', *Strategic Management Journal*, 22(4), 359–79.

Szulanski, Gabriel (1996) 'Exploring Internal Stickiness: Impediments to the Transfer of Best Practice within the Firm', *Strategic Management Journal*, 17(Winter Special Issue), 27–43.

Tsai, Wenpin and Sumantra Ghoshal (1998) 'Social Capital and Value Creation: The Role of Intrafirm Networks', *Academy of Management Journal*, 41(4), 464–76.

Tsoukas, Haridimos (2002) 'Introduction: Knowledge-based Perspectives on Organizations: Situated Knowledge, Novelty, and Communities of Practice', *Management Learning*, 33(4), 419–26.

Vincent, Leslie H., Sundar Bharadwaj and Goutam N. Challagalla (2003) 'Does Innovation Mediate Firm Performance? A Meta-analysis of Determinants and Consequences of Organizational Innovation', Working paper, The Technological Innovation: Generating Economic Results, 1–33.

von Hippel, Eric (1988) *The Sources of Innovation* (Cambridge, Mass.: MIT Press).

Yamin, Mo and Juliet Otto (2004) 'Patterns of Knowledge Flows and MNE Innovative Performance', *Journal of International Management*, 10(2), 239–58.

Zander, Udo and Bruce Kogut (1995) 'Knowledge and the Speed of the Transfer and Imitation of Organizational Capabilities: An Empirical Test', *Organization Science*, 6(1), 76–92.

Part II

International Businesses, Local Market Interactions and Impacts

4

Institutional Influences on Global Marketing Strategies

Fred van Eenennaam, Claudia M. L. Janssen and Keith D. Brouthers

Introduction

As firms expand abroad they need to make a number of important strategic decisions. Among these is whether to use a global marketing strategy or adapt their marketing strategy to the specific requirements of each country. We develop a theory that suggests the distance in different components of the institutional environment influences the standardization decision for different dimensions of marketing strategy. Our findings suggest that the heterogeneity of international institutional environments has a significant but differential impact on marketing strategy standardization and adaptation. This implies that successful international expansion may require adaptation of some marketing strategy components while standardizing others.

Recent reductions in international barriers to trade and advances in technology have led many to suggest that multinational firms may benefit from pursuing global strategies (Aaker and Joachimsthaler, 1999; Zou and Cavusgil, 2002). Sandy Lawrence, vice-president of global marketing at Polaroid stated this clearly: 'The world is global and you can't win in a marketplace unless you have a global approach' (Dietrich, 1999, p. 4). Nancy Wiese, director of worldwide marketing communications for Xerox tends to agree, stating, 'Business is conducted pretty much the same everywhere so there are opportunities for global messages' (Dietrich, 1999, p. 5).

The benefits of using global strategies may arise from scale economies, global brand value perceptions, or the growing homogeneity of consumer demand. Global strategies entail, among other things, the standardization of marketing processes, programmes and branding (Johansson and Yip, 1994; Samiee and Roth, 1992). Despite the potential benefits, multinationals such as advertising agency WPP Group tend to take a more flexible approach to global standardization. As the WPP Group chief executive officer (CEO) and founder, Sir Martin Sorrell, stated, 'One size doesn't fit all. Consumers are more interesting for their differences rather than their similarities' (White and Trachtenberg, 2003, p. B1). Moreover, research examining the performance implications of global standardization of corporate marketing strategies has provided conflicting and confusing results (Cavusgil and Zou, 1994; Johansson and Yip, 1994; Samiee and Roth, 1992; Szymanski *et al.* 1993; Zou and Cavusgil, 2002).

One potential reason for these conflicting recommendations and findings is that the anxiety of making international business decisions is compounded by four types of international business risk. Cavusgil (2006) suggests that these risks include commercial, currency/financial, political/legal, and cross-cultural: what has been referred to as the institutional environment of a country (Scott, 1995). These institutional differences may compound the confusion over firm's strategies in foreign markets, especially if these markets are distinct (institutionally) from what the firm is use to.

We hypothesize that differences in institutional environments influence the potential success of using standardized global marketing strategies. More specifically, we suggest that, because of the heterogeneity of country-level institutional environments, global standardization of marketing strategy elements may not be the best solution. Instead, we maintain that managers should make the standardization/adaptation decision based on the institutional distance between countries (Kostova, 1999). Our theory rests on two important insights. First, firm-specific advantages may be context specific, and may not be equally appropriate and applicable in every institutional setting (Brouthers, 2002; Kostova and Roth, 2002; Xu and Shenkar, 2002). Second, the institutional environment is multi-dimensional and formed from at least three specific domains – regulative, normative and cognitive – that vary between countries (Scott, 1995).

We develop a theory to suggest the circumstances under which institutional pressures from each of the three domains of the institutional environment drives firms' decision-making regarding the degree of standardization or adaptation for specific components of their marketing strategy: processes, programmes and brand image. We suggest that the distance between countries' regulatory institutional environment influences marketing process and programme strategy standardization decisions. Further, we suggest that normative institutional distance influences both programme and branding strategy standardization decisions. Finally, we suggest that cognitive institutional distance influences branding strategy standardization decisions. We test these relationships on a group of multinational corporations operating in Western Europe.

Institutional theory and standardization

The success or failure of marketing strategy standardization may be influenced by the institutional environment (Iyer, 1997). The institutional environment is important because it sets the 'rules of the game in a society' (North, 1990, p. 3). Firms that are proficient at playing by these rules tend to perform better than firms that cannot adapt to institutional demands (Oliver, 1997). Because these rules tend to be specific to each national setting, strategies that work in one country may not work as well in another, where the rules may be different (Kostova, 1999). Strategic decisions regarding compliance or defiance of institutional rules can have a significant impact on firm performance (Oliver, 1997), thus it is important for firms to identify these rules and develop strategies that provide for potential success.

Because each nation has its own institutional environment, firms investing in foreign countries may be confronted by either similar institutional environments or by very different ones. The difference in institutional environments between countries has been termed 'institutional distance' (Kostova, 1999; Xu and Shenkar, 2002). Kostova (1999) defines institutional distance as the difference in institutional dimensions – regulative, normative and cognitive – that might have an impact on the acceptability of organizational practices. In developing firm strategy for a foreign market, firms need to consider the institutional distance and how it might influence their ability to implement existing strategies effectively (Kostova, 1999; Xu and Shenkar, 2002) and exploit firm-specific advantages (Kostova and Roth, 2002).

Here we theorize how the regulative, cognitive and normative institutional environment (Scott, 1995) may influence three aspects of international marketing strategy: process strategies, programme strategies and brand strategies. Marketing process strategies are concerned with how the marketing campaign is developed and presented, how marketing activities are monitored and evaluated, and how marketing is planned (Kreutzer, 1988; Zou and Cavusgil, 2002). Marketing programme strategies are concerned with the marketing mix used in a particular country, and include product, pricing and distribution programmes (Kreutzer, 1988; Zou and Cavusgil, 2002). Brand strategies deal with brand image and brand value issues (Roth, 1995a; 1995b).

Regulatory influences

The regulatory institutional environment is composed of the laws and rules that govern business in a particular country (North, 1990; Scott, 1995). The institutional regulatory environment encapsulates laws dealing with business operations and, from a business perspective, the rules governing employee behaviour and distribution. Host market regulatory institutional factors tend to enter organizations through the people they employ (Kostova and Roth, 2002; Scott, 1995). Actions undertaken by these people are guided by habit and rules developed within a specific regulatory environment (Iyer, 1997).

When firms have developed effective marketing processes and programmes, they prefer to extend those processes and programmes to new markets (Samiee and Roth, 1992). Yet institutional theory suggests that a firm's success may depend on instituting marketing strategies that are isomorphic with the host market regulatory institutional environment (Oliver 1997). Because marketing processes and programmes are internal to the firm, they need to comply with the rules and habits of the employees of the organization. This suggests that the processes and programmes used may be standardized if the regulatory institutional environment is similar between countries (Xu and Shenkar, 2002).

Applying standardized processes and programmes, developed in one market, can help to provide efficient and effective implementation of the marketing strategy in other markets (Kreutzer, 1988; Yip, 1989). But standardized processes and programmes need to comply with (i) host government rules and regulations; as well as (ii) employee and professional standards (Handelman and Arnold, 1999). When regulatory environments are similar, employees tend to perceive these processes and programmes as being legitimate and consistent with their own regulatory institutional environment, which leads to employee compliance and support (Kostova, 1999).

When the regulatory environments differ between countries, marketing process and programme adaptation may be needed to contend with differences in legal requirements and employee standards (Xu and Shenkar, 2002). Institutional theory suggests that strategies that do not conform to the institutional norms of the host market may not be viewed as legitimate (Kostova and Zaheer, 1999). The lack of compliance with government-based regulatory institutions may result, for example, in the loss of property rights, on imposition of taxes, tariffs or duties (Iyer, 1997). The lack of legitimacy in programmes and processes may also create resistance from employees (Kostova, 1999). This may occur because new programmes and processes differ from those embedded in employee rules and habits (Iyer, 1997). Legitimacy may be achieved by adapting the firm's marketing programme and process strategies so that they are perceived as isomorphic with the foreign market regulatory environment, by both government and employee groups. Thus, to be successful, firms need to balance the benefits of standardization (such as economies of scale) with the costs of not being isomorphic with the host market's regulatory environment.

The regulatory institutional environment may have less of an impact on brand strategies because brands and brand image are the cognitions and perceptions of consumers; they are not methods or processes for doing business (Keller, 1993; Low and Lamb, 2000). Country-of-origin research suggests that consumers from different countries perceive products differently (Kotler and Gertner, 2002; Peterson and Jolibert, 1995). This is true also for different brands (Aaker and Joachimsthaler, 1999; Alashban *et al.*, 2002; Hsieh, 2002; Roth, 1995a, 1995b). The country-of-origin literature tends to suggest that brand perceptions are based on national values and beliefs held by consumers in a host country, and that these values and beliefs have a significant impact on consumer behaviour. Yet consumer values and beliefs are not part of the regulatory institutional environment. Hence, the regulatory institutional environment may have a significant impact on marketing programmes and processes but little impact on brand strategies, because regulatory institutions concern themselves with the way business is enacted, as opposed to what consumers believe and perceive.

> *Hypothesis 1a*   **When regulatory institutional distance is small, firms prefer standardized marketing process strategies. As the distance increases, so does the degree of process strategy adaptation.**

> *Hypothesis 1b*   **When regulatory institutional distance is small, firms prefer standardized marketing programme strategies. As the distance increases, so does the degree of programme strategy adaptation.**

Normative influences

The normative pillar of the institutional environment deals with procedural or behavioral legitimacy (Grewal and Dharwadkar, 2002). The normative dimension is concerned with the way people think about the world around them and how they react in certain situations (North, 1990; Scott, 1995). The normative environment provides a prescriptive, evaluative and obligatory dimension to an individual's actions (Baum and Oliver, 1991). Norms specify how things should be done; and they define legitimate actions/behaviours. Normative institutional influences designate ways of pursuing goals and objectives (Scott, 1995). Normative beliefs may influence the buying and psychological behaviour of consumers, because they impose constraints and obligations on behaviour (Bello *et al.*, 2004).

People in different countries develop different normative behaviours; for example, normative behaviour associated with ethical acts vary from one country to another (Roxas and Stoneback, 1997). In some countries, child labour is considered ethical, while in others it is seen as unethical (Kolk and van Tulder, 2002). These behavioural differences lead to different employment activities and varying consumer reactions towards organizations in different countries (Kolk and van Tulder, 2002; Roxas and Stoneback, 1997). Gaining legitimacy may entail the adoption of country-specific normative beliefs (Grewal and Dharwadkar, 2002).

Because marketing processes are internal to the firm, they may not be influenced by differences in normative behaviour. Marketing process strategies are concerned with how the marketing campaign is developed, monitored and planned (Kreutzer, 1988; Zou and Cavusgil, 2002). These internal functions may be influenced by regulatory institutional dimensions (as discussed above), but tend not be influenced by the external normative behaviour of consumers.

In contrast to this, marketing programme strategies may be affected by normative institutional differences, because they are concerned with product, pricing and distribution, which have a direct impact on consumer behaviour (Kreutzer, 1988; Zou and Cavusgil, 2002). Foreign firms may have difficulty in convincing consumers to buy a product simply because some element of it (size, shape, use or price) appears inconsistent with local behaviours. For example, Wal-Mart is having some difficulty in Japan, because consumers there think cheap prices mean poor quality. It is a challenge for Wal-Mart to convince customers that top-quality goods can be sold at lower prices (Zimmerman and Fackler, 2003).

Marketing programme features that provide a competitive advantage in one country may not be as effective in another. This suggests that, to achieve success, firms can use standardized marketing programmes in normatively close countries but must adapt their marketing strategies to provide a better fit with the existing institutional environment when the normative distance between countries is large. Hence it appears that normative institutional differences may have a significant impact on the standardization/adaptation of marketing programme strategy.

Brand strategies may also be influenced by differences in normative environments. Brand strategies are related to the image of the firm's brand (Roth, 1995a, 1995b). Country-of-origin research suggests that consumers from different countries perceive brands differently, which leads to differences in consumer behaviour (Aaker and Joachimsthaler, 1999; Alashban *et al.*, 2002; Hsieh, 2002; Roth, 1995a, 1995b). Hence the normative institutional environment influences, at least in part, consumer behaviour towards brand strategy. In normatively similar institutional settings, a standardized brand strategy may lead to desirable (consistent) consumer behaviour, while increased normative institutional differences may lead to other, less desirable, behaviour, suggesting that brand strategy may need to be adapted in these cases.

Hypothesis 2a   **When normative institutional distance is small, firms prefer standardized marketing programme strategies. As the distance increases, so does the degree of programme strategy adaptation.**

Hypothesis 2b   **When normative institutional distance is small, firms tend to use standardized brand strategies. As distance increases so does the degree of brand strategy adaptation.**

Cognitive influences

The cognitive–cultural pillar of the institutional environment has an impact on the meaning people attribute to words, signs, gestures and so on (Scott, 1995). The cognitive environment provides 'subconsciously accepted rules and customs', which in turn influence 'moral codes, expectations of trust and reliability' (Bruton *et al.* 2002, p. 205). The cognitive component of the institutional environment influences the stereotypical view that members of a society hold that influences 'the way a particular phenomenon is categorized and interpreted' (Kostova and Roth, 2002, p. 217).

The cognitive institutional environment may have a significant influence on brand strategy standardization. This tends to be the case because brand strategy (brand image) is a cognitive perception by consumers (Hsieh *et al.*, 2004; Low and Lamb, 2000; Keller, 1993). Brand strategy deals with building brand value, which can act as a competitive advantage for a firm (Alashban *et al.*, 2002; Low and Lamb, 2000; Roth 1995a, 1995b). As Xu and Shenkar (2002) suggest, differences in cognitive beliefs may create barriers to the successful internationalization of intangible advantages, such as branding. Cognitive differences may create greater or lesser value for certain brands in one institutional setting than in another (Alashban *et al.*, 2002; Hsieh, 2002; Roth 1995a, 1995b).

When institutional environments are similar, consumers may share a similar interpretation of brand image and value (Alashban *et al.*, 2002; Hsieh, 2002; Hsieh *et al.*, 2004; Roth, 1995a, 1995b). In this case, standardized brand strategies may lead to successful brand image and value creation in foreign markets (Hsieh 2002; Roth 1995a, 1995b). Standardized brand strategies may be isomorphic with the values and beliefs of foreign consumer groups. However, when cognitive institutional environments are distant, brand strategies will need to be adapted to be successful. Because of differences in consumer perceptions and values, similar brand strategies may lead to different interpretations and undesirable results (Aaker and Joachimsthaler, 1999; Alashban *et al.*, 2002; Hsieh, 2002; Roth, 1995a, 1995b). Products and services conceived as being helpful and beneficial in one institutional context may have the opposite image in another. For example, in the USA, the Disney Corporation has been very successful with its amusement parks. However, in France, Euro Disney has not fared as well. Part of the problem has been attributed to Disney's corporate image, which is perceived by part of the French population as 'American cultural imperialist, a destroyer of a French pastoral setting, and a totalitarian system impervious to the rights of employees as individuals' (Forman, 1998, p. 253). Hence, as cognitive institutional distance increases, firms may find that adaptive brand strategies lead to more successful international operations.

In contrast to this, the cognitive institutional environment may have little impact on marketing processes and programme strategies, because these factors are governed by the laws and rules of doing business in a given country and by the normative behaviour within a societal group, and not by the beliefs and values of consumers. Governments set laws about what, where, when and how certain products or services can be advertised, promoted and distributed. For example, some countries prohibit the advertising of alcoholic beverages, while others allow product comparisons; some allow television advertising, while others do not. Rules and laws concerning business processes such as accounting and record keeping also vary by country, forcing firms to adopt different processes in different regulatory institutional settings. Consumer behaviour influences the need to present marketing programmes that conform to the norms of each location. Legitimizing the actions of foreign firms requires conformity with normative values in target markets. Hence the cognitive component of the institutional environment may have little impact on marketing processes that are concerned with the rules and laws by which business is enacted (the regulative environment), or marketing programmes that are influenced by the behaviour of consumers (the normative environment).

Hypothesis 3 **When cognitive institutional environmental distances are small, firms prefer standardized brand strategies. As the distance increases, so does the degree of brand strategy adaptation.**

Methodology

Data for this study came from business units operating in consumer products industries in Western Europe. As in Zou and Cavusgil (2002), the business unit was our unit of analysis. To qualify each business unit had to be active in at least four European countries and had been marketing in the European Union (EU) for at least two years. A total of 1,306 business units, from both consumer-product and pharmaceutical (over the counter – OTC) industries were identified across seventeen countries of Western Europe, with the aid of Europages, the Amadeus database, the Dutch 'Stichting Merkartikel Association', and the Fortune-1000 list.

Each business unit was contacted by telephone, to obtain the name of the international marketing manager or director responsible for Europe, then all managers responsible for Europe were sent a letter telling them to expect the questionnaire to arrive within a week. The questionnaire was accompanied by an introductory letter explaining the nature and content of the research. All companies received the questionnaire in English, with the exception of the companies located in France. The questionnaire was translated from English into French by French native speakers and pre-tested by French native speakers with business experience. Three mailings were made to all non-responding companies.

A total of 272 questionnaires (a response rate of 21 per cent) were returned, of which 137 were usable (the remaining 135 questionnaires were returned incomplete). In total, the effective response rate was 10.5 per cent (137/1306), which is a typical response rate for a pan-European study (Harzing, 1997). A t-test was used in order to assess non-response bias. The results showed no significant differences between early and late respondents (Armstrong and Overton, 1977).

Dependent variables

We examined five dependent variables in the study, each measuring the degree of standardization/adaptation for a specific multi-dimensional aspect of marketing strategy. Factor analysis was used to verify the content of each of these dimensions. For each of our dependent variables, a high value was related to greater adaptation.

First, we examined two marketing process strategies: general marketing processes and sales processes (Kreutzer, 1988). General marketing processes consisted of seven Likert-type questions measuring the standardization of marketing planning, marketing communications, new product development, marketing control, marketing research, marketing information systems, and marketing human resource processes (Cronbach alpha=0.88). Sales process standardization consisted of three measures looking at sales implementation, service implementation and distribution implementation processes (Alpha=0.88).

Second, we examined three frequently discussed dimensions of marketing programme strategy: product, pricing and distribution. Product programme strategies were measured with two Likert-type questions asking about the degree of standardization of product characteristics and packaging (Alpha=0.72). We included five Likert-type items that examined the degree of standardization of pricing (consumer prices and customer prices) and distribution (types of intermediary, types of outlet and roles of sales force) programmes. Factor analysis indicated that these five questions all loaded on one variable, which we termed pricing/distribution programmes (Alpha=0.82).

Third, we examined the degree of standardization/adaptation of the firms' brand image strategy. Based on the work of Keller (1993) and Low and Lamb (2000), brand image standardization was measured with three Likert-type questions examining similarities in consumer perceptions of brand image, brand quality and brand benefits (Alpha=0.87).

Independent and control variables

We included six independent variables, representing three dimensions of institutional distance – two regulatory, two normative and two cognitive. Each institutional distance independent variable was measured using multiple Likert-type questions, where a high value was related to larger institutional distance. The component measures for each variable were confirmed with factor analysis.

Regulatory institutional distance included legal distance and distribution channel distance. Legal distance was measured with two Likert-type questions that asked about similarity in laws (patent, antitrust, trademark and so on) and business restrictions between markets (Alpha=0.80). Distribution channel distance was measured with three Likert-type questions concerning similarity in the types of distribution channels, degree of access to distribution channels, and market coverage of distribution channels (Alpha=0.83).

We developed two measures of normative institutional distance focusing on the behavioural attitudes of consumers: buyer purchasing behaviour and psychological behaviour. Buyer purchasing behaviour was measured with seven Likert-type questions examining differences in purchasing patterns, shop loyalty, brand loyalty, product usage patterns, service requirements, product satisfaction and product disposition (Alpha=0.89). Buyer psychological behaviour was measured with five items examining the differences in demand for information; media usage; key product benefits/attributes sought; psychological product meaning, and risk perception of purchase (Alpha=0.82).

Finally, we developed two measures of cognitive institutional distance that captured the similarities/differences in buyer values and buyer beliefs between markets. Buyer values were measured with three Likert-type questions; these looked at the perceived cultural distance regarding consumer needs, expectations and opinions (Alpha=0.82), while buyer belief distance was captured with a single Likert-type question that examined perceived consumer sensitivity to country of origin.

We included three industry control variables as well as control variables for firm size and international experience. Each industry control variable represented firms in: (i) fast-moving consumer goods (FMCG); (ii) pharmaceutical OTC (Pharmacy); or (iii) consumer electronics (Electronics) industries. For each industry control variable, a value of 1 was assigned if the firm was in the industry and a value of 0 if it was not in the designated industry. Firm size was measured as the log of worldwide business unit sales. International experience was measured as the number of international countries in which the firm did business worldwide.

Results

Prior to running the regression tests, a correlation matrix was constructed. Included in the correlation matrix are descriptive statistics for each variable. As can be seen from Table 4.1, there was substantial variability and relatedness in our measures. Because of the potential for multicollinearity, we examined variance inflation factor (VIF) scores in each of our regression analyses. VIF scores were below a value of 2 in every analysis, substantially lower than the value of 10 that is commonly reflective of multicollinearity problems (Hair *et al.*, 1995).

Table 4.2 shows the results of our five regression analyses. The first two models examine marketing process strategy standardization/adaptation. We

Table 4.1 Correlation matrix

	Mean	SD	1	2	3	4	5	6	7	8	9	10	11
1 Distribution	11.8	3.4	–										
2 Legal	7.2	2.7	0.29*	–									
3 Purchasing	27.3	6.7	0.32*	0.16	–								
4 Psychology	19.5	4.8	0.23*	0.29*	0.52*	–							
5 Values	11.9	3.3	0.25*	0.15	0.55*	0.43*	–						
6 Beliefs	3.2	1.4	0.14	0.14	0.11	0.22*	0.14	–					
7 Electronics	0.11	0.31	–0.16	0.06	–0.08	–0.08	–0.10	–0.04	–				
8 Pharmacy	0.14	0.35	0.04	–0.13	0.02	–0.03	0.10	–0.18*	–0.14	–			
9 FMCG	0.60	0.49	0.06	0.06	–0.07	0.02	–0.06	0.18*	–0.42*	–0.48*	–		
10 Firm size	2.9	1.1	0.11	0.06	0.05	0.02	0.02	0.05	–0.14	0.17	0.08	–	
11 International experience	55.6	52.6	–0.13	0.02	0.07	0.18	0.17	0.02	–0.11	0.16	–0.06	0.42*	–
Dependent variables													
12 Sales	10.1	3.8	0.17	0.36*	0.13	0.15	0.08	0.03	0.02	–0.09	–0.13	–0.00	–0.13
13 General	14.0	5.3	0.22*	0.31*	0.20*	0.16	0.21*	0.19*	–0.10	–0.04	–0.05	0.09	0.05
14 Product	9.9	2.2	0.20*	0.13	0.29*	0.35*	0.15	–0.03	0.11	0.06	–0.14	–0.05	0.15
15 Price/distance	16.0	5.6	0.41*	0.33*	0.26*	0.22*	0.09	0.01	–0.14	–0.01	–0.02	–0.02	–0.02
16 Brand image	12.4	3.5	0.10	0.28*	0.13	0.31*	0.25*	–0.05	–0.15	0.11	0.03	–0.05	0.16

Notes: * $p < 0.05$.

Table 4.2   Regression analyses

	Processes		Programmes		Branding
	Sales	General	Price/Product	Distance	Brand image
Regulatory					
Legal	0.42**	0.49**	0.01	0.48**	0.32**
	(0.12)	(0.17)	(0.07)	(0.17)	(0.12)
Distribution	0.04	0.11	0.11*	0.45**	−0.02
	(0.10)	(0.14)	(0.06)	(0.14)	(0.10)
Normative					
Purchasing behaviour	0.01	0.05	0.05	0.12	−0.06
	(0.06)	(0.08)	(0.03)	(0.08)	(0.06)
Psychological behaviour	0.04	−0.06	0.12**	0.06	0.18**
	(0.08)	(0.11)	(0.05)	(0.11)	(0.07)
Cognitive					
Values	0.00	0.16	−0.06	−0.20	0.17*
	(0.11)	(0.16)	(0.07)	(0.16)	(0.11)
Beliefs	0.04	−0.54	0.15	0.29	0.36*
	(0.23)	(0.32)	(0.13)	(0.32)	(0.22)
Control variables					
Electronics	−1.9	−3.2*	1.11	−3.1*	1.45*
	(1.2)	(1.7)	(0.71)	(1.7)	(1.1)
Pharmacy	−2.2*	−2.2	0.52	−1.2	−1.08
	(1.1)	(1.6)	(0.66)	(1.6)	(1.1)
FMCG	−2.4**	−2.4*	−0.03	−1.5	0.60
	(0.86)	(1.2)	(0.50)	(1.2)	(0.81)
International experience	−0.01	0.00	0.01*	0.00	0.00
	(0.01)	(0.01)	(0.00)	(0.01)	(0.01)
Firm size	0.23	0.38	−0.25	−0.39	−0.50
	(0.34)	(0.49)	(0.20)	(0.49)	(0.33)
Constant	7.46**	6.47*	6.17**	10.5**	8.07**
R-square	0.19	0.18	0.20	0.24	0.22
F	2.62**	2.44**	2.83**	3.56**	3.24**

Notes: * $p < 0.05$; ** $p < 0.01$ (one-tail tests), n = 137; standard error in parentheses.

found that the legal regulatory institutional distance measure was associated significantly with the degree of standardization/adaptation for both sales process strategy and general marketing process strategy. We found that greater institutional distance in legal environments was significantly related ($p < 0.01$) to increased adaptation of sales process strategies. Second, we found that increased legal institutional distance was significantly ($p < 0.01$) related to increased general marketing process strategy adaptation. We did not find any significant association between the distance in the distribution regulatory environment and standardization/adaptation of sales or general marketing process strategies. Our results also indicated that firms in FMCG industries tended to adapt their sales process and general marketing process strategies, while firms in Electronics industries tended to adapt their general marketing process strategies and firms in Pharmaceutical industries tended to adapt their sales process strategies. These findings provide partial support for Hypothesis 1a.

Models three and four examined the standardization or adaptation of product and pricing/distribution programme strategies. We found that regulatory institutional distance measures (both legal and distribution) were associated significantly with pricing/distribution programme strategies, but only the institutional distance in distribution was associated significantly with product programme strategies. We found that increased regulatory distance in distribution and legal environments were related significantly to increased pricing/distribution programme adaptation ($p < 0.01$). The regression analysis examining product programme strategy showed that product programme strategy adaptation was influenced by the regulatory distance in distribution environment ($p < 0.05$) but not the legal environment distance. We also found that increased psychological behaviour distance was significantly ($p < 0.01$) related to increased product programme adaptation. In addition, we found that firms in Electronics industries tended to adapt their pricing/distribution programme strategies, and that international experience was associated with adaptation of product strategies. Hence models 3 and 4 provide partial support for Hypothesis 1b and Hypothesis 2a.

Finally, our results concerning brand image strategy provided strong support for Hypothesis 3 and partial support for Hypothesis 2b. We found that the degree of standardization/adaptation of brand image strategy was associated significantly with both our cognitive and normative institutional distance measures. Greater institutional distance in buyer values ($p < 0.05$), in normative beliefs ($p < 0.05$) and psychological behaviour ($p < 0.01$) was related significantly to greater brand image strategy adaptation. We also found that brand strategy adaptation was associated significantly ($p < 0.01$) with the distance in legal environments, and that firms from Electronics industries ($p < 0.05$) tended to adapt their brand image strategies.

Discussion, limitations and implications

In general, we found support for our theoretical suggestion that differences in institutional environmental dimensions influence the degree of standardization of marketing strategies. More specifically, we found that differences in the regulatory environment had a significant impact on the degree of (i) the sales process; (ii) the general marketing process; (iii) the pricing/distribution programme; and (iv) product programme strategy standardization. It appears that firms may benefit from the advantages of standardizing their sales process, general marketing process and pricing/distribution programme strategies when the distance in the legal institutional environment is small, but that adaptation of these components of marketing strategy should be used when the distance in the environment increases. Further, firms may standardize their product programme and pricing/distribution programme strategies if the distance in the distribution regulatory environment is small, but need to adapt these programmes when the distribution regulatory environmental distance increases.

In contrast to our theory, the distance in regulatory legal environment was associated with brand image strategy standardization. This may have occurred because of differences in intellectual property rights between countries. Without legal protection, these property rights may not be enforceable, and other firms may copy the brand, thus reducing its significance or possibly damaging a firm's reputation (through lower quality standards) (Onkvisit and Shaw, 1988). Hence, difference in legal environments may influence brand image standardization simply because of differences in legal protection.

We also found some support for the influence of the normative institutional environment on marketing strategy standardization. Firms tended to standardize their marketing product programmes and brand image strategies when the distance in psychological behaviour was small. When the distance was large, firms tended to adapt their strategies in these two areas. Yet we found no significant influence for our second normative dimension – purchasing behaviour. Further research will be needed to determine the efficacy of our results in this area.

Finally, as theorized, we found that the distance in cognitive institutional environment was related to brand image strategy standardization. We found that firms tended to adapt their brand image strategy when foreign markets were more distant in consumer values and beliefs. These findings tend to provide additional support for those scholars who suggest that brand strategy is influenced by consumer cognition, beliefs and values (for example, Keller, 1993; Low and Lamb, 2000).

In summary, our finds provide support for the idea that the distance in institutional environmental dimensions is related to the degree of marketing strategy component standardization. It appears that different dimensions of the institutional environment are related to different components of market strategy. Firm managers need to be aware of these differentiated effects so that they can make more informed marketing strategy standardization decisions.

Limitations and implications

This chapter suffers from a number of limitations. First, we examined European based business units, hence our findings may not be generalizable to firms operating in other markets. Second, although we had an acceptable response rate, we had a relatively small usable sample of active European business units. While our response rates were consistent with previous cross-national data collection efforts, future studies may find other techniques that would increase response rates for these business unit managers. Third, we examined only five dimensions of marketing strategy. Future studies may wish to examine other dimensions, such as advertising or structure. We developed six measures of institutional distance. Future studies may develop other measures or help to refine our measures of the dimensions of the institutional environment.

Despite some limitations, our study has important implications for researchers and managers of international (global) firms. From a research perspective, our study helps by contributing to the strategy standardization literature. It also helps researchers to gain a better understanding of how firms overcome the tension between the parent firm's desire for capability standardization (cost minimization) and the subsidiary's need for target market legitimacy (local responsiveness). What we suggested, and found, was that subsidiaries were able to extend parent firm advantages in markets where the institutional environment was similar, but that adaptation of certain dimensions of parent capabilities was required as institutional distance increased. Researchers can build on these findings and those of Kostova and Roth (2002), by examining how best to transfer capabilities from parent to subsidiaries when the subsidiary faces issues of legitimacy because of the distance in institutional environment between markets.

For managers, our results suggest that understanding the regulatory institutional environment across foreign markets is particularly important when determining the degree of adaptation of marketing process, programme and brand image strategies. Within the EU, as further harmonization of legislation occurs, firms may benefit from a more standardized approach to these components of marketing strategy. However, operating currently under an assumption of homogeneous values, beliefs and behaviour across countries within Europe may prove to be unhelpful. Claims of treating Europe as one homogeneous regional market will require the testing of institutional environmental differences for specific business units.

References

Aaker, D. A. and E. Joachimsthaler (1999) 'The Lure of Global Branding', *Harvard Business Review*, November/December, 137–44.

Alashban, A. A., L. A. Hayes, G. M. Zinkhan and A. L. Balazs (2002) 'International Brand-name Standardization/Adaptation: Antecedents and Consequences', *Journal of International Marketing*, 10(3), 22–48.

Armstrong, S. J. and T. S. Overton (1977) 'Estimating Non-response Bias in Mail Surveys', *Journal of Marketing Research*, 14 (August), 396–402.

Baum, J. A. C. and C. Oliver (1991) 'Institutional Linkages and Organizational Mortality', *Administrative Science Quarterly*, 36(2), 187–219.

Bello, D. C., R. Lohtia and V. Sangtani (2004) 'An Institutional Analysis of Supply Chain Innovations in Global Marketing Channels', *Industrial Marketing Management*, 33, 57–64.

Brouthers, K. D. (2002) 'Institutional, Cultural and Transaction Cost Influences on Entry Mode Choice and Performance', *Journal of International Business Studies*, 33(2), 203–21.

Bruton, G. D., D. Ahlstrom and K. Singh (2002) 'The Impact of the Institutional Environment on the Venture Capital Industry in Singapore', *Venture Capital*, 4(3), 197–218.

Cavusgil, S. T. (2006) 'International Business in the Age of Anxiety: Company Risks' (Aib-UK keynote speech) (Online). Available at: http://www.mbs.ac.uk/research/ international-business/aib-conference/documents/keynote-Cavusgil-AIB-UK2006.pdf [accessed 13 May 2006].

Cavusgil, S. T. and S. Zou (1994) 'Marketing Strategy–Performance Relationship: An Investigation of the Empirical Link in Export Market Ventures', *Journal of Marketing*, 58(January), 1–21.

Dietrich, J. (1999) 'U.S. Multinationals', *Advertising Age*, 14 June, 1–22.

Forman, J. (1998) 'Corporate Image and the Establishment of Euro Disney: Mickey Mouse and the French Press', *Technical Communication Quarterly*, 7(3), 247–58.

Grewal, R. and R. Dharwadkar (2002) 'The Role of the Institutional Environment in Marketing Channels', *Journal of Marketing*, 66(July), 82–97.

Hair, J. F., Jr., R. E. Anderson, R. L. Tatham and W. C. Black (1995) *Multivariate Data Analysis* (Englewood Cliffs, NJ: Prentice-Hall).

Handelman, J. M. and S. J. Arnold (1999) 'The Role of Marketing Actions with a Social Dimension: Appeals to the Institutional Environment', *Journal of Marketing*, 63(July), 33–48.

Harzing, A. W. (1997) 'Response Rates in International Mail Surveys: Results of a 22-country Study', *International Business Review*, 6(6), 641–64.

Hsieh, M. H. (2002) 'Identifying Brand Image Dimensionality and Measuring the Degree of Brand Globalization: A Cross-national Study', *Journal of International Marketing*, 10, 46–67.

Hsieh, M. H., S. L. Pan and R. Setiono (2004) 'Product-, Corporate- and Country-image Dimensions and Purchase Behavior: A Multicountry Analysis,' *Journal of the Academy of Marketing Science*, 32(3), 251–70.

Iyer, G. R. (1997) 'Comparative Marketing: An Interdisciplinary Framework for Institutional Analysis', *Journal of International Business Studies*, 28(3), 531–61.

Johansson, J. K. and G. S. Yip (1994) 'Exploiting Globalization Potential: U.S. and Japanese Strategies', *Strategic Management Journal*, 15, 579–601.

Keller, K. L. (1993) 'Conceptualizing, Measuring, and Managing Customer-based Brand Equity', *Journal of Marketing*, 57(1), 1–22.

Kolk, A. and R. van Tulder (2002) 'Child Labor and Multinational Conduct: A Comparison of International Business and Stakeholder Codes,' *Journal of Business Ethics*, 36(3), 291–301.

Kostova, T. (1999) 'Transnational Transfer of Strategic Organizational Practices: A Contextual Perspective', *Academy of Management Review*, 24(2), 308–24.

Kostova, T. and K. Roth (2002) 'Adoption of an Organizational Practice by Subsidiaries of Multinational Corporations: Institutional and Relational Effects', *Academy of Management Journal*, 45(2), 215–33.

Kostova, T. and S. Zaheer (1999) 'Organizational Legitimacy under Conditions of Complexity: The Case of the Multinational Enterprise', *Academy of Management Review*, 24(1), 64–81.

Kotler, P. and D. Gertner (2002) 'Country as Brand, Product and Beyond: A Place Marketing and Brand Management Perspective', *Brand Management*, 9(4–5), 249–61.

Kreutzer, R. T. (1988) 'Marketing-mix Standardisation: An Integrated Approach in Global Marketing', *European Journal of Marketing*, 22(10), 19–30.

Low, G. S. and C. W. Lamb (2000) 'The Measurement and Dimensionality of Brand Associations', *Journal of Product and Brand Management*, 9(6), 350–68.

North, D. C. (1990) *Institutions, Institutional Change and Economic Performance* (Cambridge University Press).

Oliver, C. (1997) 'The Influence of Institutional and Task Environment Relationships on Organizational Performance: The Canadian Construction Industry', *Journal of Management Studies*, 34(1), 99–124.

Onkvisit, S. and J. J. Shaw (1988) 'The International Dimension of Branding: Strategic Considerations and Decisions', *International Marketing Review*, 6(3), 22–34.

Peterson, R. A. and A. J. P. Jolibert (1995) 'A Meta-analysis of Country-of-origin Effects', *Journal of International Business Studies*, (4), 883–900.

Roth, M. S. (1995a) 'The Effects of Culture and Socioeconomics on the Performance of Global Brand Image Strategies', *Journal of Marketing Research*, 32(May), 163–75.

Roth, M. S. (1995b) 'Effects of Global Market Conditions on Brand Image Customization and Brand Performance', *Journal of Advertising*, 24(4), 55–72.

Roxas, M. L. and J. Y. Stoneback (1997) 'An Investigation of the Ethical Decision-making Process across Varying Cultures', *The International Journal of Accounting*, 32(4), 503–35.

Samiee, S. and K. Roth (1992) 'The Influence of Global Marketing Standardization on Performance', *Journal of Marketing*, 56(April), 1–17.

Scott, W. R. (1995) *Institutions and Organizations* (Thousand Oaks, Calif.: Sage).

Szymanski, D. M., S. G. Bharadwaj and P. R. Varadarajan (1993) 'Standardization versus Adaptation of International Marketing Strategy: An Empirical Investigation', *Journal of Marketing*, 57(October), 1–17.

White, E. and J. A. Trachtenberg (2003) 'One Size Doesn't Fit All', *Wall Street Journal*, (October 1), B1–2.

Xu, D. and O. Shenkar (2002) 'Institutional Distance and the Multinational Enterprise', *Academy of Management Review*, 27(4), 608–18.

Yip, G. S. (1989) 'Global Strategy . . . in a World of Nations?', *Sloan Management Review*, Fall, 29–41.

Zimmerman, A. and M. Fackler (2003) 'Wal-Mart's Foray into Japan Spurs a Retail Upheaval', *Wall Street Journal.com*. (19 September), 1.

Zou, S. and S. T. Cavusgil (2002) 'The GMS: A Broad Conceptualization of Global Marketing and Its Effect on Firm Performance', *Journal of Marketing*, 66(October), 40–56.

5

Service Multinationals and Forward Linkages with Client Firms: The Case of IT Outsourcing in Argentina and Brazil

*Marcela Miozzo and Damian P. Grimshaw**

Introduction

Until recently, service multinationals were regarded as being less capable than manufacturing multinationals of providing advanced technologies and linkages to domestic firms. But this perception has now changed, and multinationals in services are now seen as transferring new technologies – if this is defined broadly to include organizational, managerial and information processing/analysis skills and knowledge (Dunning, 1989) – and as improving the availability of competitive services inputs, helping domestic firms to become competitive internationally (Markusen *et al.*, 2005). In particular, the expansion of international outsourcing and subcontracting of business processes has contributed to the growth of a new kind of (especially US) multinational that supplies services to other firms, much like contract manufacturers in manufacturing (UNCTAD, 2004; p. 157). This is true in the call centre industry, with leading firms Convergys, ICT Group, Sitel and Sykes, but also in the IT services sector, with leading firms IBM Global Services, EDS and Accenture.

This chapter identifies the effects for client firms in (middle income) less developed countries (LDCs) of outsourcing business functions to these new kinds of global service suppliers. It investigates the forward linkage effects in Argentina and Brazil, exploring the potential for complex IT outsourcing contracts to generate technology and performance improvements in the client firms. The chapter begins with a survey of the literature on service multinationals and linkage effects on clients.

* We gratefully acknowledge support from the British Academy award LRG-37268.

Service multinationals: linkages, absorptive capacity and mobility

Many studies argue that the entry of large multinational firms to LDCs may bring technology and productivity improvements to domestic firms. In general, the focus is on backward linkages, and the literature identifies the transfer of technology from multinational buyers to upstream suppliers. This may include assistance in setting up production facilities for prospective (domestic and foreign) suppliers, the diffusion of knowledge and skills to assist in upgrading technological and managerial capabilities, assistance in the purchase of raw materials and intermediaries, and support in the market diversification of domestic suppliers (Lall, 1980; UNCTC, 1981; Watanabe, 1983). The improvement in the performance of suppliers can have positive indirect spillovers for the host economy: demonstration effects, mobility of trained labour, spin-offs, competition effects and an increase in the local integration of multinationals (Blomstrom and Kokko, 1998; Kugler, 2001; MacDougall, 1960). Nevertheless, linkages with multinationals can also have negative effects. Multinational affiliates may form linkages in protected industries, with few incentives to upgrade technologically; or may become involved in anti-competitive practices and enforce unfair terms and conditions on suppliers in oligopsonistic markets; or transfer pressure on to supplier firms where terms of employment and remuneration may be less formalized; or appropriate markets from domestic producers (Aitken and Harrison, 1999). Although the literature mentions the growing importance of forward linkages, especially with local distributors and sales organizations (McAleese and McDonald, 1978), there is little detailed assessment of the importance and effect of linkages on clients. Nevertheless, lessons from the literature on backward linkages may be relevant. One important lesson from this literature is that the ability of firms to internalize spillovers is dependent on their own absorptive capacity (Cohen and Levinthal, 1989; Lapan and Bardhan, 1973). Indeed, many authors stress the importance of the characteristics of the domestic firms to assimilate these spillovers (Blomstrom *et al.*, 2001; Kinoshita, 2001; Kugler, 2001).

Recent work on the relationship between multinationals and the evolution of the domestic software and computer services sector in countries that are 'latecomers' in this sector – particularly in India, Israel and Ireland – shows that multinationals' spillovers are not automatic, that they are influenced by country-specific factors, and depend on the absorptive capacity of domestic firms (Giarratana *et al.*, 2004; Patibandla and Petersen, 2002). In India, the domestic software sector developed in parallel with the entry of multinationals. The exit of IBM in 1977 opened the way for the entry of other multinationals that established alliances with domestic firms, paving the way for the development of the export of services (Heeks, 1996). Investment in domestic organizational capabilities enabled the move from the transportation of software professionals to work overseas on low-level programming and maintenance of customer sites (body-shopping or on-site service model) during the 1980s to offshoring in the early 2000s (Athreye, 2005). The establishment of Texas Instruments' R&D lab for chip design and development of chip-related software in Bangalore in 1985 marked the start of a new and successful R&D offshore model, which generated important demonstration effects for domestic firms, through, for example, CMM (Capability Maturity Model) certification of domestic firms (Patibandla and Petersen, 2002). In Ireland, in contrast, a combination of fiscal incentives, proximity to the European market and an English-speaking, educated population attracted multinationals (Coe, 1999). Multinationals use Ireland as an export platform, subcontracting low-value-added, low-skilled manufacturing activities such as the porting of legacy products on new platforms, disk duplication, localization (text translation, changing formats) for the European market, and assembly and packaging (Arora *et al.*, 2004; Coe, 1999; Tallon and Kraemer, 1999). Multinationals contributed to the development of a domestic industry by supplying skills and reputation, but, except for a few successful firms, the majority of domestic firms have not captured the potential of positive linkages with multinationals in terms of developing technological and marketing capabilities (Giarratana *et al.*, 2004). In contrast to these two cases, in Israel, the software industry originated independently from multinationals. An important computer hardware sector, the military complex, relations with local universities and venture capital encouraged the development of both multinationals and local firms (Breznitz, 2005; Teubal *et al.*, 2000). Unlike Ireland and India, domestic firms account for a large share of software exports, especially of security and anti-virus software. Overall, while much attention has been focused on the impact of multinationals on the production and export strategies of the domestic software and computer services industry in these countries, less attention has been paid to the impact on clients (and ultimately on the host economy) of their relationship to a small number of global IT services suppliers that dominate the world markets.

An important contribution to our understanding of the roles and capabilities of different actors involved in international production is the work on global production networks (Gereffi and Korzeniewicz, 1994; Gereffi *et al.*, 2005). This combines an analysis of core–periphery relations within the framework of multinationals and their networks of production relationships. Gereffi (1994) differentiates between 'producer-driven' and 'buyer-driven' commodity chains. Producer-driven chains characterize industries with scale economies, such as cars, computers and semiconductors, and are driven by multinationals that may outsource production but keep R&D and final goods production within the firm. Power is concentrated in the headquarters and flows downwards through the dispersed subsidiaries and value-added flows back up the commodity chain. Buyer-driven commodity chains, in contrast, characterize mainly consumer durables such as clothing, footwear and toys. The global commodity chain is driven by large retailers, who do not undertake manufacturing themselves, but may be involved in design and marketing. Corporate power originates with the retailer/brand holder and, while it can be dispersed by the independent owner-ship of manufacturers, value-added stems from the branding and marketing functions. The strength of the global production networks literature is its explicit focus on cross-national forms of economic organization and the influence of different actors on the trajectory of production. However, it pays little attention to how national institutional and economic conditions continue to exert a signi-ficant influence on the international structure of economic activities. While the important role of services in the viability of global production networks has been widely recognized (Rabach and Kim, 1994), less attention has been paid to the role of large and powerful multinationals that are global suppliers of interme-diate services – partly reflecting their focus on the manufacturing sector. In sum, therefore, both the literature on linkages and on global production networks neglect attention to (i) services (and particularly powerful multinational suppliers of services); and (ii) the consequences for clients, rather than suppliers, or sales and distribution organizations, of their market-based interactions with powerful multinational suppliers.

A final key concern with regard to multinationals and linkages is their mobile nature. Mobility of capital grants multinationals advantages over states: it increases the multinationals' range of strategic options, enabling them to rank locations hierarchically, and invest more or carry out higher-value-added operations in some locations rather than in others; it allows multinationals to exercise credible threats to pull out if their demands are not met; it diminishes the power of state sanctions over firms; it transfers the burden of providing proper infrastructure and facilities to regional and national governments; and it enables multinationals to move to areas of the world where there are weaker unions and labour legislation, allowing them to keep their labour costs down, often at the expense of worker rights (Chesnais, 1992; Thomas, 1997). There is some evidence that foreign investment in services is more 'footloose' than in manufacturing because of lower capital intensity and sunk costs as well as weaker linkages to domestic suppliers (UNCTAD, 2004). There is also evidence of a major increase in the global relocation of service jobs, away from higher-cost areas of the world to those areas where wage levels are lower, and where employment conditions and employment rights may be poorer. It is the offshoring of IT-enabled services, particularly to India, that is receiving much attention. This process is being compared to the relocation of manufacturing jobs to China, but it has been pointed out that the relocation of services can occur at a faster pace because the 'objects that are being relocated are pixels and electronic pulses that can be transmitted by photons and radio waves' rather than requiring the closing of factories and setting up of new ones (Dossani and Kenney, 2004, p. 40).

In summary, the literature on linkage effects of multinationals suggests three issues relevant to understanding the relationship between powerful multinational IT services suppliers dominating the world computer services market and the possible effects through IT outsourcing for clients and less-developed host countries. First, forward linkages with multinationals may bring technology and performance improvements to clients; however, the trigger for these linkages and their eventual nature depends to some extent on particular host country institutional and economic conditions. Second, positive linkage effects are not automatic and are contingent on the ability of clients to internalize these improvements – through their absorptive capacity. Third, the mobility of service multinationals, particularly because of the ease of relocation of services, may bring a particular character to the linkages with clients.

Research method

This chapter draws on a comparative research project carried out during 2004–5, involving case studies of three large IT services multinationals – IBM, EDS and Accenture – and the investigation of ten IT outsourcing contracts of these firms in Argentina and Brazil (see Table 5.1). For each contract, interviews were conducted with one or more senior managers from both the IT firm and the client firm (twenty-seven interviews in total). The approach adopted for the interviews with senior managers was based on in-depth, semi-structured interviews, chosen for its capacity to facilitate iterative, exploratory discovery between data collection and analysis (Yin, 1994). In addition, background data were collected through meetings with representatives of trade associations and domestic IT services firms in both countries, as well as secondary data on the three IT services firms and on the computer services sector in general from company reports, industry consultants and government reports.

The literature on linkages has explored the relationship of multinationals with domestic firms. Here we explore the linkages of multinationals with client firms with different types of ownership, including domestic firms, subsidiaries of foreign firms and privatized state-owned firms acquired by, or merged with, a domestic or foreign firm. The rationale is that the character of 'domestic firms' in Argentina and Brazil has been radically transformed by important changes in ownership, especially since the 1990s, following privatization and an important influx of foreign direct investment (FDI), particularly cross-border mergers and acquisitions. This research can therefore shed light on the consequences of the merger and acquisition boom, in terms of linkages with both domestic firms and firms that have been subjected to mergers and acquisitions and their role in economic restructuring in the host economy.

Table 5.1 Details of the ten IT outsourcing contracts investigated

IT supplier	Client	Sector	Ownership	Start year	IT services provided	Contract duration (years)	Initial value (USD millions)
IBM (ARG)	Telefónica	Communications	Privatized/ acquired by foreign firm	2000	Infrastructure and 50% of its applications	6.5	252
	BankBoston	Banking	Foreign subsidiary	2003	Infrastructure (management of platforms)	10	41
EDS (ARG)	Techint	Steel tubes manufacture	Domestic	2001	Helpdesk, applications development	Hourly priced contract	40 (total value to date)
	Renault Argentina	Automotive	Acquired by foreign firm	1993	Infrastructure and applications development and management	6	43 (initial) 100 (renewed)
Accenture (ARG)	YPF/Repsol	Oil and gas	Privatized/ acquired by foreign firm	2001	Helpdesk and applications management	5 2	n.a.
	Telefónica	Communications	Privatized/ acquired by foreign firm	2000	Applications management	5	30
IBM (BR)	Varig	Airline	Domestic	1997	Infrastructure, mainframe operations, mid-range operations, distributed environments (desktop support), applications development and maintenance	8	200

Table 5.1 (Continued)

IT supplier	Client	Sector	Ownership	Start year	IT services provided	Contract duration (years)	Initial value (USD millions)
	BankBoston	Banking	Foreign subsidiary	2004	Operation of datacentre, infrastructure, operations and support but only for mainframe and mid-range operations, no distributed desktop	10	90
EDS (BR)	Telefónica	Communications	Privatized/ acquired by foreign firm	2001	Infrastructure	5	400
	ABN Amro (Banco Real)	Banking	Foreign subsidiary	1996 1998 2003	Data processing and applications	4	120 (most recent contract)

Forward linkages with clients: the empirical evidence

This section examines the evidence concerning linkages between IT multinationals and clients focusing on three issues identified in the review of the literature in the first section of the chapter:

  (i) the particular nature of forward linkages with IT multinationals;
 (ii) the role of client absorptive capacity in benefiting from linkages; and
(iii) the effect on linkages of the location decisions of multinationals and clients with respect to IT outsourcing.

(i) Particular nature of forward linkages with IT multinationals

Forward linkages are associated with pecuniary externalities (Hirschman, 1958), which take place through market transactions. The argument is that clients derive increased productivity from using that particular input rather than others that are less specialized and less appropriate to the specific needs of the firm (Alfaro and Rodriguez-Clare, 2004). In line with this claim, the client firms included in this study (both domestic and foreign-owned) were motivated in their decision to outsource by the anticipated advantages of obtaining a package of world-class IT services, based on common processes and tools as applied anywhere in the world by these oligopolistic suppliers (Gereffi *et al.*, 2005).

Many of the linkage effects stem from the fact that IT outsourcing is accompanied by transformation in the wider production system of the client firm (see Miozzo and Grimshaw, 2005). But the triggers to outsource IT and the nature of linkage effects take particular forms in Argentina and Brazil because of the radical policies of liberalization and privatization, the rapid increase in product market competition during the 1990s (especially from multinationals), mergers and acquisitions by foreign multinationals, and the economic context involving prolonged periods of severe recession.

An important linkage effect for clients flowed from the capability of IT services firms to implement IT systems on a common platform across the different branches and offices of the client firm (see Table 5.2). In two of these cases, the trigger to outsource IT was the acquisition by a European firm of a newly privatized client firm in Argentina and Brazil (cases IBM1 and EDS7). The European firm inherited entrenched organizational structures and legacy IT systems, and in response used IT outsourcing as a means of creating a common corporation approach with standard IT infrastructure and applications:

> An important shift in the second half of the nineties and a great demand for services came from firms that arrived to Argentina as global firms... They think of global applications and not on developing local ones, and of alignment with headquarters. (IBM1, IT firm manager, AR)

Table 5.2   Changes in the client's production system accompanying IT outsourcing

Case	Main changes in client's production system
IBM1 AR	Development of a common IT infrastructure and applications across different subsidiaries of the client following privatization and acquisition by a foreign firm
IBM2 AR	Restructuring of IT budget from fixed costs to 'pay-for-use' variable costs, as well as improved security
IBM3 BR	Improvements in a number of business processes, including reservations, finances, engineering maintenance and crew scheduling
IBM4 BR	Improved ability to offer similar services as a large organization yet remain specialized and small
EDS5 AR	Reductions in the time to process bids with over 200 suppliers through a carefully designed intranet system
EDS6 AR	Rationalization and avoidance of legal problems with dealings with a large number of smaller subcontractors
EDS7 BR	Development of common IT infrastructure and applications across different subsidiaries of client following privatization and acquisition by foreign firm
EDS8 BR	Creation of new line of activity, including card processing, applications and call centre and financial processes
Accenture9 AR	Transformation programme included the design and implementation of electronic and physical networks, which allowed central monitoring of all elements of the sales operations and the use of benchmark data to keep prices competitive, the new common platform also enabled client headquarters to change electronically the prices of the product sold (up to four times a day), in response to conditions within micro-markets.
Accenture10 AR	Improvement in a range of application management functions, including commercial sales management, product sales management, logistics, service billing and human resources

The motivations for a common platform were different in the case of Accenture9 AR, but nevertheless were believed to have generated significant cost savings. Faced with a new competitive challenge as a result of privatization and acquisition by a foreign firm, the client engaged with the IT multinational on a transformation programme affecting all parts of the client firm, from corporate offices to branches delivering frontline services. Changes included the design and implementation of electronic and physical networks, which allowed central monitoring of all elements of the sales operations and the use of benchmark data to keep prices competitive. The new common platform also enabled client headquarters to change electronically (up to four times a day) the prices of the product sold, in response to conditions within micro-markets.

Forward linkage effects were also observed in evidence of cost reductions and productivity improvements in some clients. This contrasted with in-house underinvestment in IT systems in the previous years because of unfavourable economic conditions in Argentina and Brazil. For example, in the case of EDS5 AR, client managers claimed to have benefited from dramatic reductions in the time taken to prepare and process bids with over 200 suppliers through a carefully designed intranet system, contributing to an estimated 30 per cent improvement in productivity.

A particularly idiosyncratic type of linkage effect was associated with the Argentine banking sector. Client managers argued that, prior to the collapse of the banking system in Argentina in 2001, banks did not consider outsourcing IT services because of security risks. After the crisis, however, banks faced severe economic problems caused by a huge increase in the number of transactions and in regulations from the Central Bank; this followed the economic measures taken in response to the 2001 crisis to stop the draining of bank accounts, including the freezing of bank accounts, limitation on withdrawals and the exchange of deposits in US dollars and large deposits in pesos into bonds. The responses from the banking sector included the downsizing of branches and the workforce, and the adoption of strategies to transform fixed costs into variable costs. IT outsourcing thus emerged as a strategy to make costs variable with changing demand. In the bank included in this research, the aim was to shift from a position where 70 per cent of in-house IT spending was fixed, to an outsourcing contract involving 30 per cent fixed and 70 per cent variable costs.

(ii) Absorptive capacity and linkages

Despite the evidence provided above of productivity and technology improvements from IT outsourcing, there are important differences in the ability of client firms to benefit from these linkages. Here we draw on Cohen and Levinthal (1989), who argue that a firm's absorptive capacity depends on prior related knowledge that confers on it the ability to recognize the value of new knowledge and to assimilate it. The concept of absorptive capacity has been operationalized in many ways, using as indicators firms' R&D intensity (Barrios and Strobl, 2002; Kinoshita, 2001), their level of technology relative to best practice (Girma, 2005) or an index combining several innovation inputs to firms (Chudnovsky *et al.*, 2004). We argue that, in the context of outsourcing of business functions, an important part of a client firm's absorptive capacity and of its ability to use prior related knowledge in the provision of these services in-house, is constituted by its expertise in the design and operation of contractual agreements. This organizational ability is a key mediator in the relationship between external knowledge (in the outsourced service) and client firm performance, enabling clients to assimilate, or absorb, the new knowledge associated with the outsourced function and to leverage this knowledge into productivity and technology improvement.

The challenge for the client firm is to use its experience in supplying these services in-house to negotiate with the IT supplier over improvements in the information systems and to assimilate the knowledge in the outsourced business service. We found significant differences in the client firms' ability to reap the potential benefits of linkages through IT outsourcing. In a number of cases, managers of client firms claimed to have benefited from the careful contract design stipulated by the IT firm. First, such contracts typically required strict monitoring and control of the quantity and quality of IT services (including penalties for under-performance). Managers at the Brazilian airline, for example, had to monitor more than 500 performance indicators, and this contractual requirement for quality checks (imposed by the IT firm) was believed to have generated improvements in the areas of reservations, ticketing, boarding, airline control, crew control and maintenance.

Second, the standard contract typically required the regular upgrading of hardware and software. In one case, the client manager explained that, prior to outsourcing, it was nearly impossible to win the argument for the regular upgrading of systems. The outsourcing contract provided a strategic solution for client IT managers by specifying annual IT investment over the contract duration (IBM3 BR). Also, the client in another case had made no upgrades in its in-house IT systems in the two years following the country's economic crisis. Under outsourcing, the IT firm offered a contract that imposed upgrading on the client:

> Without having a contract we wouldn't have been... obliged to update the technological equipment... We wouldn't have changed equipment, and in 5 or 6 years we would have had a completely obsolete IT park... But now we are imposing an annual updating of equipment, which was indeed in the contract, so that when the contract expires we have reasonable equipment and not obsolete technology. (IBM1, client, AR)

There were cases where the clients had made special investments to step up their contracting expertise prior to signing the agreement. For example, at IBM2 AR and IBM4 BR, where the client was in fact the same multinational operating in each country, the client manager in Brazil recollected that the company drew on a range of advice on legal, technical, financial and purchasing issues provided by a consulting firm, and sought help in the drafting of the contract from the IT director and in-house lawyer at the client HQ offices in the USA.

In all the cases discussed above, careful attention to the design, management and administration of the contract helped to improve service delivery (in

Table 5.3 Client approaches to contracting

	Client firm directs contract design and operation	Client develops a multi-supplier contract	Client firm weak influence on contract design and operation	Client chooses a 'bodyshopping' contract
Case with a foreign-owned client	IBM2 AR IBM4 BR	IBM1 AR EDS7 BR Accenture10 AR	EDS8 BR	EDS5 AR Accenture9 AR
Case with a domestic-owned client			IBM3 BR	EDS6 AR

many cases over what was done previously in-house) and leverage value added across the wider production system of the client. Moreover, in some cases, their prior expertise helped them to avoid one of the pitfalls of outsourcing to a large oligopolistic supplier–lock-in (Lacity and Willcocks, 2001)–through multi-supplier contracts (see Table 5.3). The three cases where a multi-supplier contract was developed, all involved the same client firm–a multinational with operations in both Argentina and Brazil. This client outsourced IT infrastructure in its Brazilian operations to EDS, and in its Argentinian operations to IBM. Moreover, in both countries, it outsourced half its IT applications to IBM and the other half to Accenture. The client thus opted for a multi-supplier strategy as a 'counterweight' (client firm manager, AR) to avoid becoming locked in with a single supplier.

In contrast to the above, in two cases, clients admitted that they failed to use their accumulated expertise in providing the service in-house to design and operate a sophisticated contract that could secure technology and performance improvements (EDS8 BR and IBM3 BR). At IBM3 BR, the client was very slow in resolving a contractual issue over the extent to which the IT firm, or the client, was responsible for provision of appropriate infrastructure, such as buildings and air conditioning, and human resources to deal with changing circumstances. Its weak expertise in contract negotiations led to three years of protracted discussion, obstructing service improvements during this period.

There are three cases, however, where despite clients having the necessary prior experience in the supply of services in-house, and the ability to design and operate IT outsourcing contracts, they had opted for less sophisticated 'bodyshopping' contracts in response to pressures for cost control in an unstable economic environment (see Table 5.3). The case of EDS6 AR is illustrative. As with the other two cases, client managers were especially concerned to manage the IT services through highly detailed terms and conditions to control costs. The contract provided a minimum guarantee of work for forty people with sixty days notification of any change or 'exit clause'. Based on a detailed list of professional IT workers, the client made requests based on an hourly rate for a specific level of skill. As such, no overall value was agreed with the IT supplier. IT firm managers argued that the client's reasons for contracting this way were (i) the client regarded itself as a South-American-based firm subject to great uncertainty and unable to predefine all the IT support it needed in a long-term plan; and (ii) not sufficiently mature to handle complex contracting with the IT supplier. Representatives of the IT supplier considered this form of contracting to be problematic:

> This is a form of contract where both sides lose... What we would like to happen is that we stop counting specific hours... yours or mine... to saying 'for this package of hours, for this application support, I charge you a fixed price and I guarantee certain productivity per contract'. (EDS6, IT firm, AR)

It seems that potential mutual benefits were lost at the expense of cost control. But it is notable that all three cases of bodyshopping were found in Argentina. Clearly, in a context like the Argentine economy in the years immediately after the economic crisis, it is no surprise that client firms suddenly became very interested in cost control. At Accenture9 AR, the client's contracting strategy quickly changed following the acquisition of the newly privatized client firm by a Spanish multinational. Before outsourcing, it operated 'like an American firm that outsourced many activities' (Accenture9, IT firm, AR), subsequently it was reluctant to sign long-term contracts and switched to using time sheets for greater control. Also, at EDS5 AR, the use of bodyshopping was implemented as a transition strategy towards gradual in-sourcing of IT services, again following the acquisition of the Argentinian-owned client by a foreign multinational. In all these cases, client firm managers argued that such contracts traded greater performance improvements against a degree of control over costs in an unstable economic environment.

The above suggests that there are significant differences in the mechanisms, especially the preparation, design and operation of contracts, that client firms put in place to benefit from linkages with IT service multinationals. Furthermore, unfavourable economic conditions present an additional obstacle to leveraging positive linkage effects.

(iii) Global location decisions of IT multinationals and linkages

Linkages are also dependent on the global strategies of multinationals. Dunning (1993) and Zimny and Mallampally (2002), argue that service multinationals pursue integrated global organizational strategies in the same way as manufacturing multinationals, breaking up services into components that can be produced where it is more efficient, or assigning one or more affiliates a global (or regional) mandate each to provide a particular service or function to other affiliates, and to the parent firm. The three IT service multinationals pursue global strategies. The new IBM software factory in Argentina supplies 25 per cent of its maintenance and development of applications to clients located abroad, especially in the USA. Also, EDS follows what it calls a 'best shore' strategy, with the most convenient locations – defined as lower cost and more appropriate skills – as recipients of work from other locations. While this only involved thirty to forty workers in Argentina at the time of the interviews, this strategy was growing, with a new contract to provide applications for the USA and Europe, which would lead to the hiring of a further 100 workers. Moreover, EDS classified its staff into three levels: level 1 included basic operators who earn lower salaries and tend to be hired in the host country; level 2 included data centre operators working on UNIX, mainframes and Windows NT, whose work can be done at a distance; and level 3 included high-level experts, whose very specialized work is best done at a distance in skilled and lower-cost locations. EDS representatives who were interviewed argued that the firm was moving into a complex integration strategy in South America, with Argentina focusing on a centre of experts in UNIX and AS 400 IBM and Brazil on mainframes, Linux and communications. Similarly, Accenture representatives argued that it had an important income from offshore services to US and European clients in 2002. Therefore, although the strategy of the different multinationals varies, there is evidence in all cases of a global strategy:

> The strategy of IBM and EDS are different, IBM is physically located almost everywhere, and EDS not necessarily...EDS functions with what it calls 'anchor accounts'. If it obtains an outsourcing contract, it automatically establishes offices in the country and builds from there. It grows through having a client. (IBM 2, client, AR)

There are two main implications from our findings on the global strategies of IT service multinationals that have detrimental effects on linkages. First, despite evidence of a global, intra-firm division of labour, the interviews revealed a number of obstacles regarding the potential integration into the complex strategies of multinationals of affiliates in LDCs, in particular those in countries with macroeconomic uncertainty. For example, despite the fact that Accenture Argentina could offer high skills and lower salaries for the provision of certain specialized services to Spain, instability in the Argentine economy was regarded as a barrier to establishing a policy for offshoring work there.

Second, the research findings reveal that IT multinationals are able to move between countries not only their own operations, but also the execution of their contracts with clients. This raises an important issue, as these practices weaken linkage effects with clients and the host economy, since suppliers relocate their clients' outsourcing from provision within a domestic economy to provision outside it. Indeed, evidence from this research confirms the view of foreign investment in services as 'footloose' (UNCTAD, 2004), since multi-national IT services providers are able to move their operations from one subsidiary to another, especially during periods of economic downturn. For example, in the years following the crisis, EDS Argentina signed important contracts with Chilean firms including LanChile airline, Banco de Chile bank and a Coca-Cola bottler, Andina. Highly skilled and experienced employees from EDS Argentina, including the director of operations, were spending two days each a week in Chile. Also, many EDS Argentina workers were transferred from Argentina to Chile and Brazil when EDS Argentina won the contract for Coca-Cola bottling, involving Chile, Argentina and Brazil, and McDonald's, involving all Latin America, thus requiring a strategic shift in regional operations. While these migration patterns of highly skilled and experienced employees contribute to the flexibility of multinationals, it may also, in the longer term, weaken their ties with the host country, reducing the potential for positive linkage effects with clients and the host economy.

Moreover, relocation of IT services provision may also facilitate regionalization and restructuring of the business operations of (especially multinational) client firms. Indeed, as can be seen from Table 5.4, in five of the ten cases, IT outsourcing was used as a stepping stone towards the regionalization of the IT systems of the multinational client. In each of the five cases, the change was driven by the client's European headquarters. For example, when the automotive

Table 5.4   Client regionalization strategy

Strategy	Cases	Reasons for strategy
IT outsourcing as stepping stone towards international regionalization (mainly in Brazil) of multinational client	EDS5 AR	Size and specialization of Brazilian affiliate operations
	IBM1 AR, EDS7 BR	Common IT infrastructure in Latin American operations
	IBM4 BR	Failed attempt to create efficiencies through single Latin American data centre
	Accenture9 AR	Change in governance
National provision of IT outsourcing	IBM2 AR	
	IBM3 BR	
	EDS6 AR	
	EDS8 BR	
	Accenture10 AR	

firm was taken over by a European firm, the new strategy was to regionalize global operations by opening a large data centre in Brazil to handle the Mercosur market, including accounting, administration and other operations through the implementation of SAP by AtosOrigin in Brazil. This was not possible in Argentina, since EDS Argentina – which held the IT outsourcing contract – did not engage in SAP consulting. Also, since the Brazilian subsidiary had three manufacturing factories and the Argentine subsidiary had only one, a strategy to reap the benefits of economies of scale dictated that IT operations were based in the largest regional centre. The representative of the Argentine auto-motive subsidiary argued that, from the moment that decision was adopted, the Argentine subsidiary became a second-rate player in terms of IT services expenditure and decisions:

> In 2000 and 2001 with the implementation of the data centre in Curitiba we started to cut services with EDS Argentina...we started to transfer IT services to Brazil...instead of EDS Argentina, EDS Brazil bills me X a year for its service, and synergy [by centralization] with Brazil enables me to reduce costs by 15%. (EDS 5, client, AR)

As a result, there was a drop in IT expenditure in the Argentine subsidiary from USD12.2m in 2000 to USD3.6m in 2004, reflecting a massive shift in IT work to the Brazilian client subsidiary (following the collapse of auto sales after 2000). In this example, the French headquarters was driving change. Arguably, it would have faced a high risk of industrial action if it had opted to close manufacturing plants in Argentina directly. The alternative route of outsourcing IT services, and moving their provision from the Argentinian to the Brazilian subsidiary of the client and EDS was met with little attention from media or unions, despite a substantial reduction in IT employment and income in Argentina.

In two additional cases, when a European firm acquired the newly-privatized clients in Argentina and Brazil, IT outsourcing was sought as a way to facilitate plans for global restructuring and to consolidate IT services in one location:

> All [subsidiaries] were requested to have the same structure...and the same applications...To have common infrastructure would enable us to consolidate the computer centre...Also, we thought that with strategic partners like EDS or IBM – they are in all the geographic areas where [the client] was – the decision to outsource...was going to facilitate the process of having common platforms or at some time to combine to one computer centre or two in Latin America, or one in Latin America and one in Spain - one as a backup of the other. (IBM 1, client, AR).

For another client, IT outsourcing was considered to be the solution that would facilitate regionalization. The representative of the Brazilian firm argued that it was on their agenda to create a single data centre that could be outsourced:

> In 2001, we had a project on regional opportunities – how Brazil, Argentina and Chile could work together to make efficiencies. We looked specifically at datacentres. Is it possible to have a single datacentre for the region? (IBM 4, client, BR).

IT outsourcing thus emerges as a tool to facilitate flexibility and firm-wide restructuring and regionalization in the (especially multinational) client firms. One of the implications of this is that IT outsourcing has the effect of weakening linkages between the IT supplier and client firms, and separating IT and client multinationals' investment and employment decisions from host countries. This places less developed host countries at increasing risk, since their position of competitive advantage depends increasingly on multinationals' location and integration decisions, resulting from confidential global corporate strategies of upgrading, downgrading and hierarchical ranking among subsidiary production units (Chesnais, 1992).

Conclusion

This research raises new issues regarding the linkage effects of large multinational suppliers of IT services and their clients in (middle-income) LDCs. Economic and institutional features affect the nature of forward linkages to global IT suppliers (complementing the literature on backward linkages; see Blomstrom and Kokko, 1998). Also, there are significant differences in the ability of client firms to benefit from these linkages, in part contingent on their absorptive capacity (Cohen and Levinthal, 1989). The data suggest that an important component of this is represented by client expertise in the design and operation of IT outsourcing agreements. Furthermore, unfavourable economic conditions present an additional obstacle to leveraging positive linkage effects.

Nevertheless, the research also shows that IT service multinationals are able to move between countries not only their own operations, but also the execution of contracts with clients. These practices relocate clients' outsourcing from subsidiaries of suppliers initially located within a domestic economy to subsidiaries located outside it, facilitating consolidation and regionalization of business segments of the (multinational) clients (Dunning, 1993; Chesnais, 1992). The results from this research therefore not only confirm the proposition that global service suppliers are 'footloose' (UNCTAD, 2004), but show that outsourcing services to these global services suppliers may contribute to making clients (or segments of the clients' operations) themselves more 'footloose', thus weakening linkage effects. These results are all the more significant in a context of changes in ownership through privatization and multinationals' mergers and acquisitions. This raises important questions for LDCs seeking to attract and retain multinationals, as IT outsourcing brings irreversible restructuring effects to IT employment and investment in host countries.

References

Aitken, B. and A. Harrison (1999) 'Do Domestic Firms Benefit from Foreign Direct Investment? Evidence from Venezuela', *American Economic Review*, 89(3), 605–18.

Alfaro, L. and A. Rodriguez-Clare (2004) 'Multinationals and Linkages: Evidence from Latin America', *Economia*, 4(2), 113–70.

Arora, A., A. Gambardella and S. Torrisi (2004) 'In the Footsteps of Silicon Valley? Indian and Irish Software in the International Division of Labour', in T. Bresnahan and A. Gambardella (eds), *Building High Tech Regions: Silicon Valley and Beyond* (Cambridge University Press), 78–120.

Athreye, S. (2005) 'The Indian Software Industry and Its Evolving Service Capability', *Industrial and Corporate Change*, 14(3), 393–418.

Barrios, S. and E. Strobl (2002) 'Foreign Direct Investment and Productivity Spillovers: Evidence from the Spanish Experience', *Welwirtschaftliches Archiv*, 138, 459–81.

Blomstrom, M. and A. Kokko (1998) 'Multinational Corporations and Spillovers', *Journal of Economic Surveys*, 12(2), 1–31.

Blomstrom, M., S. Globerman and A. Kokko (2001) 'The Determinants of Host Country Spillovers from Foreign Direct Investment: Review and Synthesis of the Literature', in N. Pain (ed.) *Inward Investment, Technological Change and Growth* (Basingstoke: Palgrave Macmillan) 34–66.

Breznitz, D. (2005) 'Collaborative Public Spaces in a National Innovation System: A Case Study of the Israeli Military's Impact on the Software Industry', *Industry and Innovation*, 12(1), 31–64.

Chesnais, F. (1992) 'National Systems of Innovation, Foreign Direct Investment and the Operations of Multinational Enterprises', in B.-Å. Lundvall (ed.), *National Systems of Innovation* (London: Pinter), 265–95.

Chudnovsky, D., A. Lopez and G. Rossi (2004) 'Foreign Direct Investment Spillovers and the Absorption Capabilities of Domestic Firms in the Argentine Manufacturing Sector (1992–2001)', Working Paper 74, Universidad de San Andres, Buen A.

Coe, N. (1999) 'Emulating the Celtic Tiger? A Comparison of the Software Industries of Singapore and Ireland, *Singapore Journal of Tropical Geography*, 20(1), 36–55.

Cohen, W. and D. Levinthal (1989) 'Innovation and Learning: The Two Faces of R&D', *Economic Journal*, 99, 569–96.

Dossani, R. and M. Kenney (2004) 'The Next Wave of Globalization? Exploring the Relocation of Service Provision to India', *Berkeley Roundtable on the International Economy*, May 19.

Dunning, J. (1989) 'Multinational Enterprises and the Growth of Services: Some Conceptual and Theoretical Issues', *The Service Industries Journal*, 9(1), 5–39.

Dunning, J. H. (1993) *The Globalisation of Business* (London: Routledge).

Gereffi, G. (1994) 'The Organisation of Buyer-driven Global Commodity Chains: How US Retailers Shape Overseas Production Networks', in G. Gereffi and M. Korzeniewicz (eds), *Commodity Chains and Global Capitalism* (Westport, Conn.: Praeger), 95–122.

Gereffi, G. and M. Korzeniewicz (eds) (1994) *Commodity Chains and Global Capitalism* (Westport, Conn.: Praeger).

Gereffi, G., J. Humphrey and T. Sturgeon (2005) 'The Governance of Global Value Chains', *Review of International Political Economy*, 12(1), 78–104.

Giarratana, M., A. Pagano and S. Torrisi (2004) 'The Role of Multinational Firms in the Evolution of the Software Industry in India, Ireland and Israel', Paper presented at the DRUID Summer Conference, Elsinore, Denmark, 14–16 June.

Girma, S. (2005) 'Absorptive Capacity and Productivity Spillovers from FDI: A Threshold Regression Analysis', *Oxford Bulletin of Economics and Statistics*, 67(3), 281–306.

Heeks, R. (1996) *India's Software Industry: State Policy, Liberalisation and Industrial Development* (Thousand Oaks, Calif.: Sage).

Hirschman, A. (1958) *The Strategy of Economic Development* (New Haven, Coun.: Yale University Press).

Kinoshita. Y. (2001) 'R&D and Technology Spillovers through FDI: Innovation and Absorptive Capacity', CEPR Discussion paper No. 2775.

Kugler, M. (2001) 'The Diffusion of Externalities from Foreign Direct Investment', Paper prepared for RES conference, Durham.

Lall, S. (1980) 'Vertical Inter-firm Linkages in LDCs: An Empirical Study', *Oxford Bulletin of Economics and Statistics*, 42(3), 203–26.

Lacity, M. and L. Willcocks (2001) *Global Information Technology Outsourcing* (Chichester: Wiley).

Lapan, H. and P. Bardhan (1973) 'Localised Technical Progress and the Transfer of Technology and Economic Development', *Journal of Economic Theory*, 6, 585–95.

MacDougall, C. (1960) 'The Benefits and Costs of Private Investment from Abroad: A Theoretical Approach', *Economic Record*, 36, 13–35.

McAleese, D. and D. McDonald (1978) 'Employment Growth and Development of Linkages in Foreign-owned and Domestic Manufacturing Enterprises', *Oxford Bulletin of Economics and Statistics*, 40, 321–39.

Markusen, J., T. Rutherford and D. Tarr (2005) 'Trade and Direct Investment in Producer Services and the Domestic Market for Expertise', *Canadian Journal of Economics*, 38(3), 758–73.

Miozzo, M. and D. Grimshaw (2005) 'Modularity and Innovation in Knowledge-intensive Business Services: IT Outsourcing in Germany and the UK', *Research Policy*, 34(9), 1419–39.

Patibandla, M. and B. Petersen (2002) 'The Role of Transnational Corporations in the Evolution of a High-Tech Industry: The Case of India's Software Industry', *World Development*, 30(9), 1561–77.

Rabach, E. and E. M. Kim (1994) 'Where Is the Chain in Commodity Chains? The Service Sector Nexus', in G. Gereffi and M. Korzeniewicz (eds), *Commodity Chains and Global Capitalism* (Westport, Coun.: Praeger), 123–42.

Teubal, M., G. Avnimelech and G. Alon (2000) 'The Israeli Software Industry: Analysis of the Information Security Sector, TSER Project "SMEs in Europe and Asia: Competition, Collaboration and Lessons for Policy Support"', Mimeo, Hathew University, Jerusalem.

Tallon, P. and K. Kraemer (1999) 'Ireland's Coming of Age with Lessons for Developing Countries', *Journal of Global IT Management*, 3(2), 4–23.

Thomas. K. (1997) *Capital Beyond Borders: States and Firms in the Auto Industry* (New York: St. Martin's Press).

UNCTAD (2004) *World Investment Report 2004: The Shift Towards Services* (New York and Geneva: United Nations).

UNCTC (1981) *Transnational Corporation Linkages in Developing Countries: The Case of Backward Linkages via Subcontracting* (New York: United Nations).

Watanabe, S. (1983) *Technology, Marketing and Industrialisation: Linkages between Small and Large Enterprises* (New Delhi: Macmillan).

Yin, R. (1994) *Case Study Research: Design and Methods* (2nd edn) (Newbury Park, Calif.: Sage).

Zimny, Z. and P. Mallampally (2002) Internationalization of Services: Are the Modes Changing?', in M. Miozzo and I. Miles (eds), *Internationalisation, Technology and Services* (Cheltenham: Edward Elgar), 87–114.

6
Multinationals' Perceptions of their Economic and Social Impacts

Fabienne Fortanier and Ans Kolk

Introduction

In this age of anxiety, the business of business is no longer just that. The growth of multinational firms over recent decades has been accompanied by increased public concern regarding the economic, social and environmental consequences of their activities. As central actors in the economies of both developed and developing countries, the global operations of multinational enterprises are increasingly being scrutinised by governments and policy-makers, NGOs (such as trade unions or consumer organizations), and the public at large. Academics too have increased their attention to the public consequences of the private activities of multinational firms, as reflected, for example, in the debate on the role of foreign direct investment (FDI) on host-country growth and development at the macro level (see, for example, Caves, 1996; Meyer, 2004; Rodrik, 1999), and in broader contributions on the social and moral challenges of globalization (Akhter, 2004; Dunning, 2003).

As public pressure builds, MNEs are forced increasingly to respond to these concerns. Driven by their stakeholders, MNEs have started to report on their corporate principles, processes and outcomes (see Wood, 1991) with respect (primarily) to their environmental activities, as documented in studies on corporate accountability and corporate responsibility – by, for example, Doh and Guay (2004); Kolk (2005); and KPMG (2002). While such 'self-reporting' may be criticized for being an example of public relation (PR) ('window-dressing'), rather than CSR (corporate social responsibility), activities, there are several reasons for attaching importance to these statements. First, the public punishment for 'false promises' can be severe, even worse than for making no promises at all, as the well-known Nike case regarding child labour showed.[1] This induces firms to be relatively honest and realistic in what they communicate. In addition, extensive reporting is associated significantly with good environmental performance (Al-Tuwaijri *et al.*, 2004), although there are still ample possibilities for additional research in this direction (as outlined by, for example, McWilliams and Siegel, 2000). Even if the information that firms themselves provide is not entirely correct, the fact that MNEs play a large role in the world economy and are increasing expected to contribute towards dealing with societal problems (van Tulder and Kolk, 2001), already make their perceptions interesting to study and worthwhile to consider. Acknowledging potential problems is a first and necessary step towards taking action to try to solve them. In doing so, MNEs can also act as examples for other firms.

However, while corporate reporting on their environmental, and to a lesser extent ethical, practices has been analysed previously in the literature, relatively little attention has been paid to how firms perceive their socio-economic impact. At the same time, while the more macroeconomic literature analysed the role of FDI in economic growth, the mixed evidence has induced authors to propose a more detailed analysis of how MNEs influence the economic dimensions of development. This chapter aims to make a contribution towards linking these two literatures in order to understand MNEs' (potential) contribution to sustainable development. We do so by exploring how MNEs currently report on their economic and social impacts, and thus how they perceive their role of (good) corporate citizens. The chapter will explore to what extent MNEs are already aware of their social and economic impacts, and what this includes (and what it does not). We consider firms' self-reported impact of economic activities on society, employment and employee issues, and external social impacts, giving details and notable examples. Subsequently, it will be discussed how the analysis of firms' self-perception and cognition might be helpful for further research, and for managers and policy-makers interested in assessing and guiding MNE behaviour. Before moving to the impacts as reported, first the current literature is summarized briefly.

MNEs' potential impacts: literature review

The rise in worldwide FDI since the 1980s has been hailed by many as a felicitous process, for developing countries in particular. It has been considered as an important means of complementing domestic savings, transferring skills, knowledge and technology, improving competition and increasing the quantity and quality of employment; thus furthering economic growth and social development. However, MNEs have also been accused of crowding out local firms, using technology that is not always appropriate for local circumstances, creating only low-wage jobs, contributing to so-called 'McDonaldization', manipulating transfer prices, and (ab)using their political and economic power in host countries. In this section, we shall review the existing evidence regarding both claims, and indicate the main potential economic and social impacts of MNE activity. These are used subsequently to analyse the non-financial reports of MNEs.

Economic impacts

At the macro level, FDI can enlarge the production base of the host country (more capital and taxes; so-called size effects). By adding to the host country's savings and investments, it may enlarge the production base at a higher rate than would have been possible if a host country had had to rely on domestic sources of savings alone (Borensztein *et al.*, 1998).

But more important than the sheer size of the investments are the more indirect effects from FDI (also known as spillovers). These may occur in several ways. For example, an investment by an MNE in a local economy can stimulate competition and improve the allocation of resources, though fears are often expressed that MNEs may out-compete the local, often much smaller, firms ('crowding out') (Harrison and McMillan, 2003; Lall, 2000). Also, spillovers may result from linkages that MNE affiliates establish with local buyers and suppliers. Backward linkages with suppliers cannot raise the overall output of local supplier firms and increase their productivity and quality of their products, as MNEs might provide technical and managerial assistance or assist in purchasing raw materials and intermediary goods (Alfaro and Rodríguez-Clare (2004). Forward linkages – with buyers – could benefit local firms through products with lower prices or better quality (Aitken and Harrison, 1999).

Finally, since MNEs are frequently key actors in creating and controlling technology, their affiliates can be important vectors for spreading managerial skills and expertise regarding products or production processes – intentionally or unintentionally – to host-country firms (Romer, 1993). However, MNEs' (capital intensive) technologies may not always be appropriate for developing-country (labour intensive) contexts, with local firms facing difficulties in absorbing foreign technologies and skills (Borensztein *et al.*, 1998).

Whether the effect of FDI on economic development is on the whole positive or negative is an already much researched question. On the one hand, De Mello (1999), Sjöholm (1997b) and Xu (2000) found that foreign investors increase growth in host countries. Baldwin *et al.* (1999) showed that domestic technological progress is aided by foreign technological progress, and studies by Borensztein *et al.* (1998) and OECD (1998) also demonstrated that FDI had a larger impact on economic growth than investment by domestic firms. On the other hand, a study by Kawai (1994), using a set of Asian and Latin American countries, indicated that an increase in FDI generally had a negative effect on growth (with the exception of Singapore, Taiwan, Indonesia, the Philippines and Peru). The impact of FDI on growth has also been negative in Central Eastern European countries (see Djankov and Hoekman, 1999; Konings, 2000; Mencinger, 2003; UNECE, 2001). Studies that used industry-level rather than macroeconomic data (often focusing on productivity growth as being the equivalent of economic growth) did not yield consistent results (compare, for example, Aitken and Harrison (1999), with Anderson (2001) and Sjöholm (1997).

Social impacts

Moving to social impacts, two main dimensions can be distinguished. First, the firm-internal aspects related to employment creation, labour conditions and wages; and second, the rather more externally orientated contribution of MNEs, for example, to local social causes (philanthropy, respect of human rights, for example).

With respect to the first dimenion, MNEs are generally shown to create direct and, in particular, indirect employment (Görg, 2000). In addition, MNE affiliates on average pay higher wages in developing countries than do local firms, either because the former are more productive or they want to prevent employees from switching jobs to domestically-owned competitors (Globerman *et al.*, 1994). Skilled employees benefit most from this (Feenstra and Hanson, 1997; Lipsey and Sjöholm, 2004), thus increasing wage inequality between high-skilled and low-skilled workers in these cases. Regarding the overall effects of FDI (or FDI-induced growth) on inequality and income distribution, studies show inconclusive results or note that the relationship is difficult to establish (Bigsten and Levin, 2000; Tsai, 1995).

As to the quality of the employment (rather than its quantity of remuneration), labour (and other) standards might sometimes be less vigilantly enforced or even be lowered – for example, in cases when host-country governments compete to attract FDI. But there is little evidence for such a 'race to the bottom', or proof that employment conditions in MNEs are generally worse than in local firms (OECD, 1998). Neither is there evidence for a 'race to the top', however, although MNE's adherence to international (labour) standards, and human rights and anti-corruption conventions, might induce a 'leading edge' role.

To what extent these potential impacts at the macro level translate into actual firm perceptions and behaviours will be analysed next. This will also reveal whether MNEs mention other aspects than the ones indicated above, and/or clarify how social and economic impacts can be seen in a different light compared to when they are analysed from a macro perspective.

Sample and variable selection

To obtain information about firm's perceptions of its socio-economic impacts, we analysed the non-financial reports of the Fortune Global 250. Using this list, as published in July 2004, we collected, in the period September 2004–January 2005, their most recent corporate report, which dealt with environmental, social responsibility and/or sustainability issues. This could be either a separate report or, if this is not available, the annual financial report if it contained this kind of information. Websites were visited to search actively for reports, but if this did not yield results, the companies were contacted, several times if necessary, by letter, mail and/or phone, in order to have certainty about reporting by the whole set of 250 companies.

For economic impact, we analysed the reports on five different elements. This first involved discovering whether firms mentioned basic financial information in their report at all, and in particular related to the (distribution of) their value-added. This can be seen in the way that firms elaborate on the size of their activities, or the direct effects. Second, we identified whether firms mentioned their more indirect consequences for the countries they invest in (for example, through technology transfer and linkages). Finally, we distinguished three more focused economic topics that a selection of firms mentioned in their non-financial reports: tax issues, fair competition and fair trade.

For social impacts, we explored MNEs' perspectives regarding employees, distinguishing dimensions such as training, working conditions, equal oppor-tunities and freedom of association. To grasp the more externally-orientated contribution of MNEs, we looked at issues such as human rights and corruption, but also at corporate philanthropy and attention to HIV/AIDS, health programmes or water projects.

Social and economic impacts in MNE reports

Impact of economic activities on society

Table 6.1 gives a general overview of the economic issues that MNEs include in their reports. A distinction is made between those that mention this kind of information, either generally or more specifically (the category 'mention' in the tables); and those (the second category) that provide such more detailed information with figures and/or an explicit elaboration (column 'also specified'). Overall, it can be seen that basic information on sales, profits and other balance sheet or income statement information, is reported most often (62 per cent). Least mentioned are tax issues (tax payments, transfer pricing), and particularly

Table 6.1 Economic issues included in Fortune Global 250 CSR reports, 2004 (in percentages)

	Mentioned	Also specified
Basic information (e.g. profits)	61.49	40.37
Impact of economic activities on society	25.47	13.66
Tax issues (e.g. tax payments, transfer pricing)	15.53	2.48
Fair competition	6.21	2.48
Fair trade	6.21	3.11

fair trading practices and fair competition. A quarter of the companies address more directly the topic of the impact of their economic activities on society; almost 14 per cent also give more detailed information about this. A concrete example of such 'economic impact reporting' is Alcoa (2004, p. 48), which indicates that their Suriname subsidiary 'accounts for roughly 15% of Suriname's GDP if multiplier effects are taken into account'.

Most notable also is that European and Japanese MNEs have started to report on their 'added value' to society and stakeholders, and include figures on (cash) value added to different groups, or on how expenses are distributed. As to the categories distinguished, most firms reporting the breakdown of their value added identify some common, well-known 'stakeholders' (employees, shareholders, creditors, suppliers, government). But a range of other labels and groupings are used. For example, Société Générale reports on the distribution of value added to shareholders, governments and staff only. Some companies also include the 'income kept within the company' (Vivendi Universal); 'invested in business for future growth' (Unilever); 'retained for investment' (HSBC); also designated simply as 'business' (Telecom Italia); 'company' (BASF; Volkswagen); or 'enterprise system' (Enel) as a category in addition to the stakeholder mentioned earlier. In a few cases, there is so-called 'community investment' or 'voluntary contributions to charities and NGOs' (BP; Unilever; Ito-Yokado) also listed as one of the items.

Although the wide range of approaches prevents direct comparisons, the trend towards the specification of contributions offers insight into the different means of reporting about economic impacts on society; or, as BASF (2004, p. 57) puts it, 'Unlike the statement of income, the value added statement is not from the shareholder's perspective, but explains BASF's contribution to private and public income.'

Generally speaking, MNEs do not differentiate between individual countries when reporting on their economic impact, thus hampering an assessment of the impact of FDI. In a few cases, however, some of this information is given. Dow Chemical gives figures for salaries, taxes and purchasing per region (North America, Europe, Latin America, Pacific). Telefónica indicates, for six countries in South America, how it contributes to the economies and their development, by listing economic data (revenue in relation to GNP, number of employees, and local suppliers).

As for another element in Table 6.1, tax issues, Pemex (2004, p. 4) contributes 'more than 33% of the federal income which allows the government to cover a sizable share of the country's social expenditures'. ExxonMobil (2004, p. 18) calculated that its operations 'generated more than 70 billion dollars in taxes paid to local, state and national governments. This equals 200 million per day'. Alcoa (2004, p. 49) reports its effective tax rate, which declined from 33.5 per cent in 2000 to 24.2 per cent in 2003.

Fair trade, mentioned least frequently (together with fair competition) in the reports is generally used in the sense of offering fair-trade products such as coffee and bananas (for example in company restaurants). In Japan, fair trade (also referred to as 'fair trading') appears to be seen as synonymous with fair competition (see, for example, the Japanese Fair Trade Commission, which implements the country's competition policy). The two Japanese firms that elaborate more substantially on fair trade (Ito-Yokado and Matsushita Electric Industrial) in fact refer to fair competition. Under the heading of 'fair competition', Japanese firms also seem to be most explicit. Interestingly, they tend to refer to fair competition among their *suppliers* for contracts rather than to their own behaviour towards other competitors. For example, 'Providing fair competition: NEC will make information on procurement available in a timely and appropriate manner so as to provide fair business competition to all domestic and overseas suppliers who wish to take part in business deals' (NEC, 2004, p. 18).

Impact of MNEs on social issues: employees

If we move from the more general economic impact on society to a specific indication of social implications, MNEs frequently refer to employment and employee issues. This can be an indication of corporate size in general, such as Carrefour (2004, p. 42), 'one of the 10 largest employers in the world... with more than 430,000 employees'. Others refer to the impact in their home country: 'The total economic impact of BT in terms of income generation and employment in the UK is calculated by adding the direct, the indirect, and the induced, impacts. This totals to 8.944 GBP billion in employee income, and 431,753 employees, that account for approximately 1.7% of all jobs in the UK' (BT, 2004, p. 22). For Hewlett-Packard (HP) (2004, p. 6) 'job creation extends beyond the number of people a company employs directly. For example, the Sacramento (California) Regional Research Institute determined that 2.3 jobs are created for every HP job based in the Sacramento region'.

A few companies give indications of employment in host countries. Coca-Cola (2004, p. 16) in this regard reports on its impact in Africa: 'With nearly 60,000 employees in 54 countries and territories, the Coca-Cola system is Africa's largest private sector employer.' Sometimes, specific examples are given of employment creation and other benefits as a result of MNE activities abroad. ExxonMobil (2004, p. 20) states that its Chad–Cameroon project (which 'spent more than 780 mil US$ with 2200 local businesses' and also brought infrastructural improvements for local villages) offered employment to 3,500 people during construction, '80% of whom were citizens of Chad or Cameroon'.

Unilever and Total report not only on the size of their workforce, but also on their efforts to increase the quality of human resources through training: 'Total deploys action programs to increase the number of local employees and suppliers, usually supported by vocational training courses' (Total, 2004, p. 126). At Unilever (2004, p. 12) 'local employees are trained in advanced performance management systems, have access to world-class technology and gain experience in leading-edge consumer research and retailing strategies. Unilever Vietnam has consciously sought to develop a range of sourcing and distribution partnerships with local companies that require the transfer of technology capabilities. These partnerships now support 5,500 jobs, in addition to the company's 2,000 direct employees in Vietnam'.

Compared to these more external aspects (employment creation and societal impact), the more 'internal' aspects related to employees receive much more attention in MNE reports (as an outflow of more traditional social reporting). Table 6.2 gives an overview of topics such as health and safety, training, working conditions, equal opportunities and diversity policies. On these issues, companies provide quite a lot of information, much of it quite detailed. This is much less the case with, in particular, collective bargaining and freedom of association, where specific details are rarely offered. These two stand out for the purpose of this chapter, because they refer to international (labour) standards that MNEs can endorse (or not) through their international activities.

Although a third of the MNEs refer to the organization of employees, either via statements on the freedom to associate or by elaborating on collective bargaining processes, most do not expand on these claims and include only a few sentences, such as International Paper (2004, p. 24), for example, which states that, to comply 'with the employment laws of every country in which we operate, and we recognize lawful employee rights to freedom of association and collective bargaining. We deal directly with legally constituted labor organizations'. A company that reports very specifically on the extent of unionization is Telefónica (2004, p. 72): 'Freedom of association is a right of Telefónica employees, as shown by the fact that more than 42,000 employees are members of a trade union. The 2,584 trade union representatives chosen in these used more than 850,000 trade union hours during 2003.'

Table 6.2   Employee issues included in Fortune Global 250 CSR reports, 2004 (in percentages)

	Mentioned	Also specified
Health and safety	72.05	57.14
Training	72.05	47.83
Working conditions	62.11	23.60
Equal opportunities	60.87	32.92
Diversity	67.70	39.13
Employee satisfaction	31.68	18.01
Collective bargaining	32.92	14.29
Freedom of association	26.09	4.97

Typical of Japanese firms – and a reflection of the local institutional context – is the presence of company-specific, or company-internal 'labour unions'. The unions generally have a high number of members, sometimes even by definition (that is, all employees are automatically union members – as, for example, Tepco). The reports stress in particular the commonality of the goals of both management and labour unions: 'the company and labour union create orderly labour–management relations through sincere communication, while respecting and valuing each other's positions'. (Marubeni, 2004, p. 36).

European MNEs often mention their Works Councils. Peugeot (2004, p. 86) reports that 'European employees are represented by a European Works Council created in 1996, whose role, resources and European membership were expanded in 2003'. This also takes place at worldwide level: 'Formed by unanimous agreement on October 27, 2000, the Renault Group Works Council is the employee representative body at overall Group level. It is composed of 36 representatives from Renault's majority-owned subsidiaries in the European Union, as well as in Brazil, Argentina, Korea, Romania, Slovenia and Turkey (16 countries)' (Renault, 2004, p. 21).

Impact of MNEs on social issues: external social consequences

In addition to the labour rights mentioned above, many MNEs also pay attention to other international norms and standards. In particular, respect for human rights is mentioned frequently (in 50 per cent of the reports; see Table 6.3), but much less often specified. Almost a third of the reporting MNEs also refer to their rejection of forced/child labour, but here again details, and thus indications of implementation, are rare. This also applies to corruption/bribery, which is mentioned even less often. Overall, most MNEs leave it at a general statement of commitment to human rights and a condemnation of violations of standards, but do not further specify their policies to ensure the

Table 6.3 External social issues included in Fortune Global 250 CSR reports, 2004 (in percentages)

	Mentioned	Also specified
Human rights	50.93	10.56
Forced/child labour	29.81	4.97
Corruption/bribery	18.01	4.97
Philanthropy	73.91	30.43
Foundation	47.20	21.12
Employee involvement/volunteering	58.39	24.84
Education/school programmes	65.22	24.22
Health programmes	40.37	13.66
HIV/AIDS	28.57	14.29
Water projects	10.56	3.73

implementation of these claims. None in fact reports (with data) on human rights violations related (directly or indirectly) to their company. Some MNEs, including ING, Norsk Hydro and ABB, substantiate their commitments by engaging in partnership agreements with Amnesty International.

With regard to bribery and corruption, companies mention affiliation to the Transparency International Initiative (for example, Siemens, HSBC, Norsk Hydro, BASF). In a few cases, further information is given: 'Deutsche Telekom has introduced comprehensive measures to comply with legal requirements and to rule out corruption and misappropriation. Clear Group guidelines apply and we monitor compliance in Group-wide precautionary joint audits. Group audits regularly check the cooperation between different local audit units and our domestic and international business units' (Deutsche Telekom, 2004, p. 24).

The final way in which MNEs can have an impact on society is through community contributions. This includes employee volunteering, philanthropy, support for water, health, education and other kinds of project (see Table 6.3). Almost three-quarters of all firms mention giving to charity, either money or 'in kind', as part of their overall corporate social responsibility activities. Nearly half of the reports indicate that these philanthropic activities also occur via a separate company Foundation. Often, firms focus their charitable giving in areas related to their core business.

Deutsche Telekom (2004, p. 80), for example, has 'been supporting the crisis line "Telefonseelsorge" for many years by providing our telephone network and technical and organizational services without charge. The calls made correspond to a value of nearly €2 million'. Carrefour (2004, p. 41) opened several charity shops that 'not only provide food aid to the needy but also to integrate disadvantaged young people into the workforce by offering them vocational training in retailing'.

Employee involvement is also a way in which internal objectives (employee commitment) can sometimes be linked to external causes. Close to 60 per cent of the MNEs report that part of the way in which they shape their initiatives of community involvement includes the sponsoring and stimulation of employee volunteering. Often, quantitative information is given about the number of projects carried out, the number of employees involved or the total amount of hours/days spent on volunteering. Typical projects involve cleaning and (light) renovation works (of parks or community buildings), involvement in education and organizing activities for disabled or elderly people. Projects may be organized or initiated either by the employees themselves, or by higher-level management.

Alcoa (2004, p. 44) is a company that also uses employee volunteering for other purposes: 'When Alcoa needs to close, idle, or curtail work at one of its facilities for competitive reasons, the company works to help minimize adverse effects on the surrounding community and maximize opportunities for each Alcoan affected by the situation. For example, an Alcoa smelter located in

Wenatchee has been idle since 2001...By year's end 2003, the facility's employees had contributed more than 80,000 hours of community service.'

The 'good causes' that benefit from corporate charity are diverse. Table 6.3 shows that school and education projects are often mentioned (65 per cent of the reports). These projects can range from 'science grants' for high-school children or training programmes for graduate students in developed countries, to humanitarian programmes (building primary schools) aimed at reducing illiteracy. Charitable health programmes – specifically HIV/AIDS – are mentioned by 30 per cent of the reports. These projects are generally reported to involve a combination of education or awareness training, and free (or cheap) medication. They can be aimed at either the company's own employees, at local communities, or both. A representative example is Barclays (2004, p. 15):

> Barclays Africa is implementing a comprehensive HIV/AIDS Management programme to assist its employees to deal with the impact of the HIV/AIDS pandemic. The programme includes: a) provision of health care and support for HIV positive employees, including counselling, financial planning assistance, medical treatment and access to anti-retroviral drugs; b) education and prevention programmes for employees to understand how HIV is transmitted and why they should not discriminate against HIV positive employees; c) establishment of policies that manage sensitive issues related to HIV positive employees including commitment to continuation of employment and guarantee of confidentiality; and d) strong focus on peer educator training and involvement in the community. Barclays is currently providing anti-retroviral drugs to HIV positive employees and up to three dependants in Botswana, Zambia, South Africa and Kenya. Plans are in place to extend the programme to other countries in 2004.

Pharmaceutical companies have a special interest in the issue of HIV/AIDS. The patents on the anti-retroviral drugs are part of their key competitive advantages (and the result of large previous investment). Yet pressure from society on pharmaceutical companies to distribute medication at low cost, or allow developing country firms to copy the medication, are large. Roche (2004, p. 12) reports the following regarding its patent policy:

> To improve access to those most in need of urgent life-saving HIV/AIDS medicines, Roche has developed a specific HIV/AIDS patent policy: Roche will not file patents on new antiretroviral therapies in the Least Developed Countries and sub-Saharan Africa; Roche will not take action against generic versions of its antiretroviral therapies where Roche holds, or has licensed-in, the patent in the Least Developed Countries and in sub-Saharan Africa;

Roche holds no patents for the malaria medicines Fansidar or Lariam in the Least Developed Countries and sub-Saharan Africa.

Conclusions

This chapter set out to explore what MNEs have to say about the economic and social impact of their global activities. The large role that MNEs play in the world economy means that they are looked upon increasingly as contributors dealing with social problems. The trends toward privatization, self-regulation and public–private partnerships further illustrate that governments and societies look increasingly to (large, multinational) firms to provide public goods, and value. The self-perception of MNEs regarding the extent to which they may be part of those problems, or can be part of the solution, is therefore worthwhile to researching and considering. By examining these self-perceptions, this chapter aimed to make a contribution to both the literature on corporate social responsibility – by analysing in particular the hitherto under-researched socio-economic dimensions of CSR – and the literature on the economic consequences of inward FDI – by exploring corporate strategy at the micro, rather than macro level of analysis.

Based on a review of the literature, we distinguished several socio-economic elements of sustainable development that we used to analyse the non-financial reports (for example, CSR reports, sustainability reports) of the 250 largest firms worldwide (based on the Fortune Global 2004 list). Driven partly by institutional and stakeholder pressures, firms are increasingly disclosing information about the social, environmental and, very recently, the economic implications of their activities, in non-financial, 'triple bottom line' reports. At the moment, up to two-thirds of the firms that have a report mention one or more aspects of their socio-economic impact – representing almost half of the Fortune Global 250.

While an important part of the Fortune Global 250 firms reported in some way on their socio-economic impact, the detailed description of the contents of the reports showed that the variety of issues discussed and the methods of measurement differed enormously across firms. Still, some interesting conclusions can be drawn from the illustrations that were discussed above.

First, we found that the information provided in the reports was primarily positive, and often presented in the form of examples, rather than a systematic assessment of all activities. Reports highlight the 'best practices', or indeed 'showcases', while the potential negative consequences of MNE behaviour were not addressed. Similarly, with a few exceptions, no distinction was made regarding the impact for individual host countries.

A second striking element was the extent of quantitative information that is presented in the reports. A considerable number of firms highlight the number of people employed, directly and indirectly, the number of humanitarian projects they are involved in or the amount spent on philanthropy. While the lack of 'bad news' raises suspicions about such reports as merely providing window-dressing, or being PR activities, the precision and measurability (and hence controllability) of information that is presented in some reports suggests that certain firms have started to take their socio-economic impact seriously. In particular, the considerable number of firms that specify their value added across stakeholders may provide an interesting – and cross-firm comparable – way of operationalizing and measuring a substantial part of a firm's impact on the economies of the countries in which it has investments.

Third, the reports by firms seem to reflect the institutional and cultural background of their countries of origin, as well as demonstrating sector peculiarities. Sector differences are apparent in the nature of the philanthropic projects in which firms are engaged: pharmaceutical companies sponsor HIV/AIDS programmes, while computer and telecom firms support schooling projects aimed at bridging the digital divide, for example. Country differences are potentially even stronger: the nature of the home-country business system seems to determine both the extent and the way in which firms perceive their socio-economic consequences. For example, European and Japanese firms both appear to be quite proactive in addressing societal concerns, but have a different understanding of the problems, and a different range of solutions. Fair trade has different implications in European Union (EU) versus Japanese firms. And freedom of association for workers is dealt with differently in Japan – mainly through company-specific trade unions – than it is in Europe (where larger trade unions and Works Councils are preferred).

More quantitative research could further explore these differences across countries of origin, industries, and other firm characteristics. Such studies could also increase the detail of policy recommendations from these findings. For now, the results presented in this chapter suggest that, while governments look increasingly to MNEs to take responsibility in addressing societal and economic problems, they should probably not have too much faith in these 'market solutions', given the still limited number of firms that engage in these activities. If MNEs contribute, it is more likely to be on a project base than in a structural manner. In addition, policy-makers may want to realize that definition and interpretation differences of the various issues exist among MNEs from different cultural backgrounds, meaning that firm solutions will probably also differ, and may not be in line with policy intentions. Still, what MNEs currently have to say about their roles in social and economic issues probably presents a first step in the direction of more involvement in dealing with socio-economic issues. The details on the distribution of firms' value added could be a useful way of comparing the contribution of firms to the countries in which they invest. Further research is necessary to link MNEs' perceptions to their 'real' impacts: do firms that apparently perceive and respond to public pressure to be good citizens also translate these concerns into good practices? And do host-country circumstances moderate this relationship? Individual company case studies as well as multi-method approaches that link the information presented in this chapter to more macroeconomic data on development outcomes should shed light on these and other questions.

Note

1. Nike was a front-runner company, being one of the first to adopt a code of conduct promising not to use child labour to make its products. However, NGOs discovered that its suppliers still did so, which dealt a major blow to Nike's reputation. At the same time, other sporting goods manufacturers that did not claim publicly to have abandoned child labour operated in ways similar to Nike but were not targeted.

References

Aitken, B. and Harrison, A. (1999) 'Do Domestic Firms Benefit from Direct Foreign Investment? Evidence from Venezuela', *American Economic Review*, 89, 605–18.

Akhter, S. H. (2004) 'Is Globalization What It's Cracked Up to Be? Economic Freedom, Corruption, and Human Development', *Journal of World Business*, 39(3); 283–95.

Alcoa (2004) *2003 Sustainability Report* (Pittsburgh, Penn: Alcoa).

Alfaro, L. and A. Rodriguez-Clare (2004) 'Multinationals and Linkages, an Empirical Investigation', *Economia*, 4(2), 113–69.

Al-Tuwaijri, S., T. E. Christensen and K. E. Hughes II (2004) 'The Relations among Environmental Disclosure, Environmental Performance, and Economic Performance: A Simultaneous Equations Approach', *Accounting, Organizations and Society*, 29, 447–71.

Anderson, G. (2001) 'Spillovers from FDI and Economic Reform', NEUDC conference, Boston, Mass., 28–30 September.

Baldwin, Richard, H. Braconier and R. Forslid (1999) 'Multinationals, Endogenous Growth and Technological Spillovers: Theory and Evidence', CEPR Discussion paper No. 2155.

Barclays (2004) *Corporate Social Responsibility Report 2003* (London: Barclays).

BASF (2004) *Corporate Report 2003* (Ludwigshafen: BASF).

Bigsten, A. and J. Levin (2000) 'Growth, Income Distribution, and Poverty: A Review', Economics Working paper No. 32, Göteborg University.

Borensztein, E., J. De Gregorio and J.-W. Lee (1998) 'Does FDI Affect Economic Growth?', *Journal of International Economics*, 45, 115–35.

BP (2004) *Sustainability Report 2003. Defining Our Path* (London: BP).

BT (2004) *Social and Environmental Report* (London: BT).

Carrefour (2004) *Sustainability Report 2003. Promoting Socially Responsible Commerce and Constructive Globalisation* (Paris: Carrefour).

Caves, R. (1996) *Multinational Enterprise and Economic Analysis* (Cambridge University Press).

Coca-Cola (2004) *Towards Sustainability: Coca-Cola Citizenship Report* (Atlanta, Ga.: Coca-Cola).

Deutsche Telekom (2004) *The 2003 Human Resources and Sustainability Report* (Darmstadt: Deutsche Telekom).

De Mello, L. R. (1999) 'FDI-led Growth: Evidence from Time Series and Panel Data', *Oxford Economic Papers*, 51, 133–51.

Djankov, S. and B. Hoekman (1999) 'Foreign Investment and Productivity Growth in Czech Enterprises', *World Bank Economic Review*, 14, 49–64.

Doh, J. P. and T. R. Guay (2004) 'Globalization and Corporate Social Responsibility: How Non-governmental Organizations Influence Labor and Environmental Codes of Conduct', *Management International Review*, 44 (Special Issue), 7–29.

Donaldson, T. (1989) *The Ethics of International Business* (New York: Oxford University Press).

Dunning, J. H. (2003) *Making Globalisation Good* (Oxford University Press).

ExxonMobil (2004) *Corporate Citizenship Report* (Irving, Tx.: ExxonMobil).

Feenstra, R. and G. Hanson (1997) 'Foreign Direct Investment and Relative Wages: Evidence from Mexico's Maquiladoras', *Journal of International Economics*, 42(3/4), 371.

Globerman, S., J. Ries and I. Vertinsky (1994) 'The Economic Performance of Foreign Affiliates in Canada', *Canadian Journal of Economics*, 27, 143–56.

Görg, H. (2000) 'Multinational Companies and Indirect Employment: Measurement and Evidence', *Applied Economics*, 32, 1809–18.

Harrison, A. and M. McMillan (2003) 'Does Direct Foreign Investment Affect Domestic Credit Constraints?', *Journal of International Economics*, 61(1), 73–100.

HP (2004) *Global Citizenship Report* (Palo Alto, Calif.: Hewlett-Packard).

International Paper (2004) *International Paper Sustainability Report 2002/2003* (Memphis, Tenn.: International Paper).

Ito-Yokado (2003) *Corporate Social Responsibility Annual Report 2003. We Wish to Remain a Trusted Company Known for Our Integrity* (Tokyo: Ito-Yokado).

Kawai, H. (1994) 'International Comparative Analysis of Economic Growth: Trade Liberalisation and Productivity', *The Developing Economies*, 17(4), 373–97.

Kolk, A. (2005) 'Environmental Reporting by Multinationals from the Triad: Convergence or Divergence?', *Management International Review*, 45, 145–66.

Konings, J. (2000) 'The Effect of Direct Foreign Investment on Domestic Firms: Evidence from Firm Level Panel Data in Emerging Economies', CEPR Discussion paper No. 2586.

KPMG (2002) *KPMG International Survey of Corporate Sustainability Reporting 2002* (De Meern: KPMG).

Lall, S. (2000) 'FDI and Development: Policy and Research Issues in the Emerging Context', Working Paper 43 Queen Elizabeth House, University of Oxford.

Lipsey, R. E. and F. Sjöholm (2004) 'Foreign Direct Investment, Education and Wages in Indonesian Manufacturing', *Journal of Development Economics*, 73, 415–22.

Marubeni (2004), *CSR Report 2004*, (Tokyo: Marubeni).

McWilliams, A. and D. Siegel (2000) 'Corporate Social Responsibility and Financial Performance: Correlation or Misspecification?', *Strategic Management Journal*, 21(5), 603–9.

Meyer, K. E. (2004) Perspectives on Multinational Enterprises in Emerging Economies', *Journal of International Business Studies*, 35, 259–76.

Mencinger, J. (2003) 'Does Foreign Direct Investment Always Enhance Economic Growth?', *Kyklos*, 56(4), 491–508.

NEC (2004) *Annual CSR Report* (Tokyo: NEC).

OECD (1998) *Open Markets Matter: The Benefits of Trade and Investment Liberalisation* (Paris: OECD).

Pemex (2004) *Report on Sustainable Development 2003* (Mexico City: Pemex).

Peugeot (2004) *Annual Report and Business Review 2003* (Paris: Peugeot).

Renault (2004) *Sustainable Development Report* (Paris: Renault).

Roche (2004) *Sustainability Report 2003* (Basel: Roche).

Rodrik, D. (1999) *Trade Policy and Economic Growth: A Skeptic's Guide to Cross-national evidence*, CEPR Discussion paper No. 2143.

Romer, P. (1993) 'Ideas Gap and Object Gaps in Economic Development', *Journal of Monetary Economics*, 32, 543–73.

Sjöholm, F. (1997), 'Productivity Growth in Indonesia: The Role of Regional Characteristics and Direct Foreign Investment', *Economic Development and Cultural Changes*, 47, 559–84.

Telefónica (2004) *Corporate Responsibility Annual Report 2003* (Madrid: Telefónica).

Total (2004) *Corporate Social Responsibility Report 2003: Sharing Our Energies*, (Courbevoie: Total).

Tsai, P.-L. (1995) 'FDI and Income Inequality: Further Evidence', *World Development*, 23(3), 469–83.

Tulder, R. van and A. Kolk (2001) 'Multinationality and Corporate Ethics: Codes of Conduct in the Sporting Goods Industry', *Journal of International Business Studies*, 32(2), 267–83.

Unilever (2004) *Summary Social Review 2003. Listening, Learning, Update on Progress*, (Rotterdam/London: Unilever).

UNECE (2001) *Economic Survey of Europe, No.1* (Geneva: UNECE).

Wood, D. J. (1991) 'Corporate Social Performance Revisited', *Academy of Management Review*, 16(4), 691–718.

Xu, B. (2000) 'Multinational Enterprises, Technology Diffusion, and Host Country Productivity Growth', *Journal of Development Economics*, 62, 477–93.

Part III

Political and Strategic International Business Challenges

7

MNCs' Actions in the Socio-political Market: A Study of a Case with a Network Approach

Joong-Woo Lee, Pervez N. Ghauri and Amjad Hadjikhani

Introduction

Business network studies have explored relationships between firms and other business actors extensively but rarely touched on their socio-political relationships. In the absence of research in this area, the aim of this chapter is to address this shortcoming in business network approach and study how MNCs manage their relationship with socio-political organizations to strengthen their position in the international market. The study develops a theoretical view and stresses the firms' managerial behaviour when interacting with socio-political organizations. The framework contains the three interrelated concepts of legitimacy, commitment and trust which describe the firms' behaviour. The concepts in the model are then tested with a case study of Daewoo Motor Company, a Korean MNC in Poland and the European Union (EU) market.

Despite the fact that the interplay between socio-political organizations and business firms is an important dimension for MNCs, the study of this interaction has attracted little research (Boddewyn, 1988; Crane and Desmond, 2002; Keillor and Hult, 2004; Ring *et al.*, 1990). The increasing competition in foreign markets has forced MNCs not only to strengthen their position in the business scene, but has also affected their ability to manage socio-political actors. But researchers have been less comfortable in using complicated theoretical tools for further understanding of the MNCs' socio-political environment (Welch and Wilkinson, 2004). Authors such as Dixon (1992) and Welch and Wilkinson (2004), for example, cite the failure to incorporate socio-political issues into marketing texts. Despite a large volume of international news published in the mass media and an extensive general public discussion about the interaction between the two, such topics have seldom attracted researchers.

This study holds the view that the socio-political environment is a source of uncertainty, and that firms undertake market actions in order to eliminate or absorb that uncertainty. The presumption observes that MNCs undertake actions to influence socio-political actors. Socio-political actors, such as governments, use their legitimate power and employ coercive or supportive actions to decide and implement market rules that affect MNCs. The aim of business MNCs, on their side, is to undertake strategic actions to gain specific benefits and improve their market position. It lies in the long-term interest of MNCs and of the foreign socio-political actors to manage the socio-political environment (O'Shaughnessy, 1996). The aim of this study is to add knowledge about interaction between MNCs and socio-political units. Knowledge of MNCs' political behaviour can enhance our understanding of firms' entire market behaviour (Hadjikhani, 2000; Ring *et al.*, 1990).

The chapter will develop a theoretical view for studying MNCs acting in the socio-political market. The view has its grounds in business networks and its emphasis is on how the firms manage the socio-political organizations that eventually subsidize the business activities. The framework contains the three interrelated concepts of legitimacy, commitment and trust. The concepts of commitment activity and trust are introduced to describe firms' activity in developing their legitimate position. The notions in the model will then be tested with a case study (see Hadjikhani and Ghauri, 2001).

A short review of existing studies

Researchers in a new field put the emphasis on the social responsibility of the firms (Crane and Desmond, 2002; O'Shaughnessy, 1996), defined by others as social marketing. Social actions of the firms are directed towards the preservation of the ethics of society (O'Shaughnessy, 1996). A limitation of these studies is that they only consider interactions with social actors. In contrast to this view, but yet similar to the questions raised in this study, is the research on lobbying and the influence of business and interest groups on political actors (Andersen and Eliassen, 1996; Potters, 1992). In this research field, studies have examined pressure groups, bargaining and bribery (Bolton, 1991; Crawford, 1982; Rose-Ackerman, 1978) to gain influence that touches indirectly the core of this study. But these studies fall short of developing notions on the interdependency and relationship between business and socio-political bodies.

Marketing and management studies have paid a lot of attention to firms and their business markets, but the interdependencies between firms and socio-political organizations have attracted less interest among researchers (Ring *et al.*, 1990). The studies that have paid attention to this issue range from the presumption that management is a function of response to the political environment (Conner, 1991; Egelhoff, 1988; Ghauri and Cateora, 2006; Kogut,

1991; Korbin, 1982) to the design of coping strategies. In these contributions, the focal perspective is on the one-dimensional impact; that is, the hierarchical power of the political organizations and their impact on firms' marketing activities.

Some researchers in political science explain this interplay from the point of view of the state. These studies have been conducted in fields such as development economics (Esping-Andersen, 1985; Maddison, 1991) and regional economics (Hanf and Toonen, 1985; Nowtotny *et al.*, 1989). Business behaviour is studied by considering the economic theories that underlie the true political actions of the business actors. Government and its impact on firms is the main topic of these studies. In contrast to this approach, researches on governance (see, for example, Fligstein, 1990; Streeck, 1992) explain the failure of the political hierarchy and discuss the internal dynamics and differentiated social systems as factors that make centralized political control more difficult when regulating firms' behaviour. Following this track, some others use economic and socio-political theories to depict the problems of states in overcoming stable and consistent rules (Chaudhri and Sampson, 2000), the impact of government on competition (Ramaswamy and Renforth, 1996) and heterogeneity of the firms and industrial policy (Barros and Nilssen, 1999).

Some researchers in international business follow this track and suggest a dyadic view (Crane and Desmond, 2002). The main contribution is that actors on both sides are seen as being active in influencing each other. They also explain that the unidirectional influence, or the presumption of one-sided action of political organizations to regulate the market, suffers from an economic bias, a passive perspective. The contributions of Hadjikhani and Ghauri (2001), Ring *et al.*, (1990), Welch and Wilkinson (2004) rely on interaction between business and political actors. They do not study the social actors in the business network. A study that reflects deeply on the socio-political market and the firms' strategic actions to strengthen legitimacy is essentially absent.

A conceptual framework

This study goes beyond the mainstream in business networks and assumes that enterprises are interwoven in a network containing both business and non-business actors. MNCs are dependent on socio-political actors, and these actors are also dependent on enterprises, as MNCs make investments that affect actors in the socio-political environment (Ghauri and Holstius, 1996; Hadjikhani, 1996; Hadjikhani and Ghauri, 2001). In this study, relationships are seen as a set of socio-political and economic exchanges that interrelate different types of actors. Some have a generalized exchange, while others have a specific exchange in which the reciprocity is not necessarily achieved through any direct benefit to one actor over another, but may be achieved through an indirect benefit provided by a further actor. Socio-political actors in the local market and MNCs are interdependent and this interaction can be beneficial for both. MNCs are interdependent with socio-political actors as they can gain support or suffer hindrance to their interactions with local business actors.

Similar to the business market, the socio-political market is explained from a viewpoint of having a heterogeneous nature and firms are assumed to carry out market activities in order to gain specific support from socio-political units and thus strengthen legitimacy and their market position. While socio-political actors, like the government, put stress on the repeated procurement setting and homogeneous effect of the decisions, MNCs aim to gain heterogeneity in the gains and impacts. MNCs' investment in the socio-political market is to encourage a special relationship with socio-political actors and gain specific support. Such a view relies on a relationship explanation between MNCs and socio-political actors that goes as follows: socio-political units are dependent on MNCs because they have resources and commit investments that, in turn, affect groups such as media and the public at large, on which the socio-political actors are dependent (Hadjkhani, 2000, 1996; Hadjikhani and Ghauri, 2001; Jacobson *et al.*, 1993). MNCs create employment, and their production activities affect gross national product (GNP), which benefits socio-political actors. This leads to the explanation that these two actors are interdependent and their interplay is contingent upon a set of actors from business and socio-political systems.

Socio-political actors use their legitimate position to affect the market, and they decide and implement market rules that affect groups of firms homogeneously. The aim of business firms, on their part, is to undertake strategic actions towards governments, to gain specific support and improve their market position. In gaining further resources from socio-political actors to support business activities, the struggle is eventually to strengthen their legitimate position. The activities conducted are therefore seen in this study from the point of a firm's market legitimacy combined with commitment and trust. Legitimacy is defined as the position recognized by the surrounding actors in a business network. It is constructed on the strength and types of ties between a focal unit and others. The development, securing or maintenance of a legitimate position is emphasized with resource commitment and the trust developed in the inter-related ties. The greater the size of appropriate business and socio-political commitments, the greater can be the trust and the stronger will be the legitimate position of the firm in the market. Legitimacy is constructed based on the surrounding actors' knowledge of how a firm's performance preserves the rule of mutuality and maintains its own interests and those of others.

Dividing the market into business and socio-political units forces firms to act in two different types of connection with two different bases for legitimacy. While business legitimacy is composed of evaluation by connected suppliers, customers and so on, having business exchanges, the socio-political legitimacy relies on interaction with socio-political actors. Publicity can provide socio-political legitimacy, thus its impact is mediated through associations with commercial text. Socio-political actors also show legitimacy gain when they can manifest co-operation with MNCs that have a high reputation. They can show the gains in employment and economic prosperity needed to retain their political or social position. Firms' legitimacy is thus a recognition constructed by an accumulation of legitimacies reached in both business and socio-political markets. However, the legitimate position of the firms contains two interrelated types of legitimacy. One is gained from interaction with business and the other is reached from relationships with socio-political actors.

Both MNCs and socio-political actors struggle to devote resources and commit to activities that ultimately develop political legitimacy and benefit the involved parties – ultimately, the customers. In gaining legitimacy, MNCs can reach the customer in two ways. One is through the business market and business exchange. The other is the development of position through interaction with socio-political actors. In the latter case, the relationship is indirect and goes via firms' commitment and trust when dealing with rules established by socio-political actors. Commitment can be explained in terms of the size of investment or actions towards the counterpart alone, or towards the partner and it's connected actors (Denekamp, 1995; Scott, 1994). This can mean, for example, the establishment of a political unit in the firms' organizations or investing in lobbying organizations. Trust as the driving force and/or as an outcome of the commitment, is defined as the benevolence of the counterpart's actions towards the achievement of mutuality, which can be direct or indirect.

The variations in the degree of socio-political commitment and trust among firms explains differences in managerial behaviour (Keillor *et al.*, 1997; Lee, 1991). This can be related to an MNC's degree of internationalization in a specific country. In the internationalization process, the business commitment and trust (Johansson and Vahlne, 1990), and legitimacy (Hadjikhani, 1996), are cumulative. The degree of socio-political commitment and trust follows the process of business development, but at the same time is related to the aspect of the socio-political agenda (see Hadjikhani and Ghauri, 2001). Though, as shown in Figure 7.1, there can be variations in the degree of commitment and trust activities in different phases of internationalization.

MNCs confronting socio-political influence can take action in two different ways: *influential* or *adaptive*. Enterprises require rules and supportive measures, and socio-political organizations aim to gain legitimacy as these enterprises satisfy all the groups to whom they are responsible. The interaction between the two requires the settlement of conflicting interests, but also provides the condition for exploring options and sharing common values. They commit activities to gain trust and strengthen their positions. In this vein, the alternative

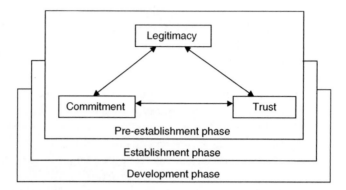

Figure 7.1   A conceptual view
Source: Based on Ghauri and Holstius (1996).

commitment options for the political actors are either *coercive* or *supportive*, and each of these contains a *general* or *specific* influence.

Where the socio-political actions are coercive, distrust prevails. Unmanageable coercive actions of socio-political units can, for example, force firms to change production units, or *threaten to exit* from the market, or resort to *bribery*. In cases where they lack the power to take such strategic actions, firms have to *adapt* or, in severe cases, file for *bankruptcy*. A highly coercive impact generates high *adaptation* cost. On the other hand, the higher the firms' resource commitment and knowledge in the business market in the relationship with political actors, the higher the influence will be and the lower the adaptation cost.

Hadjikhani and Ghauri's (2001) study shows how Swedish firms make extensive commitments to influencing political units in the EU before any political decisions are made. The greater the influence, the more heterogeneous the impact of the government and the stronger the firm's legitimate market position. Thus, *supportive* actions can be of a *specific* or a *general* nature. The firms' commitment is to generate trust and gain specific support in order to improve their market position with regard to their competitors. On the other hand, the lower the firms' commitment, the higher the adaptation will be or, in serious cases, firms will have to leave the market. The adaptation alternative is composed of new investments in the organization that will weaken the market position of the firms.

Nevertheless, the commitment options of the firms are those of *adaptation* and *influence*. Adaptation applies to the socio-political rules that enterprises have to follow. Influence means negotiation and co-operation, where firms aim at gaining specific support for their business activities. Firms with a low level of commitment to demonstrate trust rely on the first mode of political activity, which considers internal production or strategic change in order to adapt business activities to the political rules. This influence enables the firm to change the behaviour of socio-government bodies and overcome socio-political rules that specifically support the firm.

Research methodology

The study of firms' relationships with socio-political actors tends to be sensitive, and firms aim to keep the relationship out of sight of public scrutiny, being afraid of negative publicity. The study started with a survey of more than 100 firms, asking for access, but very few were willing to answer. Therefore the study focused on using a case study. The information in the case study is based on both primary and secondary data. The case reported here covered the period 1992 to late 2002. It is important to research the processes over a long period of time in order to analyse and understand the socio-political behaviour of the Korean MNC in combination with establishing and developing a business in Poland.

For the MNC, the primary information was collected via in-depth interviews with forty-two directors and managers of DMC, Daewoo Electronics and Daewoo Corporations in Korea, and their subsidiaries and R&D units in England, France, Germany and Poland, and with three Polish politicians in the Commission. Other interviews were conducted with Korean managers in Korea, and with Korean and Polish directors and managers of Daewoo-FSO (Fabryka Samochodow Osobowych), Daewoo Motor Poland (DMP) and Centrum Daewoo. The secondary data were collected from brochures and information released by these firms on their socio-political operations in Poland, and through published information provided by Daewoo-FSO and a Korean organization advising on exports – KOTRA.

The twenty-two joint venture firms (with 300–500 employees) are engaged in the production and sale of automobile components, three also produce electronics, and almost all of the firms manufacture the strategic components supplied to Daewoo-FSO and DMP. The total investment of DMC, including the joint venture (JV) firms, is approximately USD 2.3 billion. Data on these firms were collected via interviews with the chief executives, directors and managers at the level of individuals and groups (2–3 persons). The interview guide was semi-structured into the following sections: organization, production, marketing, purchasing, contact with government authorities, trade unions and other organizations as well as dealing with socio-political and cultural factors. Interviews were also held with intermediary organizations, both public (like the Euro Trade Association) and private (commercial consulting companies) in Korea. Furthermore, interviews were held with hybrid organizations, such as the Export Commission and KOTRA, in both Korea and Poland, whose function is to support small and medium-sized firms in the EU.

The case study

The Korean MNC, DMC, made large market investments in two European countries, starting decades ago. DMC's direct investments cover more than 100 countries. The aspects of DMC discussed in this chapter cover its three subsidiaries in Poland. The case presentation follows the process of entry and expansion of DMC to Poland. The first section is devoted to the pre-establishment activities (prior to the industrial commitment), and thereafter the discussion is devoted to the process of establishment and then expansion.

The establishment of subsidiaries of DMC

DMC's industrial activities have played a part in modernization of the automobile industry in Poland. Poland, as a member of EU, had a strong interest in establishing a strategic position in the automobile market. Following the dramatic changes in the political system in Poland, Daewoo was able to utilize its experiential knowledge to establish a strategic position in the Polish market. One of the major concerns that Poland faced was the creation of a balance between foreign firms and local industries. This was accommodated with political stability, industrial development and high employment in the country. This created a number of interactions between DMC and socio-political actors, not only in Poland but even among the EU's politicians. These interactions are discussed as they occurred both before and after market penetration, because of the variety of interactions in these phases.

Pre-establishment and establishment phases

The Polish automobile industry, after the period of communist government, was faced with a large number of obstacles. A number of Polish firms, particularly FSO and Fabryka Samochodow Lublin (FSL), suffered from, for example, a lack of financial, technical and marketing management resources. As a result, production capacity was decreased and the Polish company could not meet market demand at the beginning of the 1990s. Politically, the visit by the Polish president, Lech Walesa, in 1992 was a crucial factor in developing a relationship between Daewoo and the firms in the Polish automobile industry. President Walesa was interested in the Korean automobile industry and visited DMC's Bupyung manufacturing plant to observe its modern production system. He also met the chairman of the Daewoo Group, Woo-Chung Kim, to discuss the development of the automobile industry. This was a significant moment for Chairman Kim to present DMC and discuss business opportunities for co-operation with the Polish firms. President Walesa gained a positive impression from DMC's modern manufacturing plant.

The Warsaw sales office of the Daewoo Corporation was established later in 1992. Under the direction of the director, Byung-Il Chun, and with a business report received from him, DMC begin to show interest in investing in the manufacturing plant in Poland in 1993. As a result, DMC sent the vice-president, Young-Nam Wang, to investigate the possibility of developing an automobile business in Poland. However the political situation in Poland was considered unstable and too risky for business investment. This was because of the dissolution of the country's Parliament in 1993. Moreover, the president and the government party did not have absolute political power, because the government consisted of a coalition of five parties. Negotiating with the government was difficult. The market was already open to European countries. Subsequently, DMC chose Romania to become its target market in central Europe and invested a large amount of capital to establish a certain position in that market, rather than in Poland. In 1994, when questions like deregulation became a serious issue for Polish politicians, DMC realized that there was a need to get closer to the decision-makers. Chairman Kim of the Daewoo Group decided to devote financial and manpower resources to conduct political and social activities.

The Daewoo Group had also employed a former American Secretary of State, Henry Kissinger, to function as business adviser in the matter of socio-political activities. In total, the firm engaged twelve people for this function, three of whom were located in Warsaw, six in Korea, and the rest in other European countries. The task of the advisory unit in Korea and the European HQ of Daewoo Group in Germany was to assist the unit in Warsaw by, for example, providing information or by negotiating with the Polish politicians involved in the FSL and FSO's political decisions regarding privatization. DMC was concerned with areas such as the political rules for manufacturing systems, labour laws, consumer protection laws, taxes, competition law, pricing and environmental issues.

At the end of 1994, because of the increasing political role of the Polish Parliament in the automobile market, DMC decided to utilize the sales unit in Warsaw. The main task of the unit, from the beginning of the 1990s, was to collect information on strategic issues such as the privatization, modernization and liberalization of the market. These include: (i) the rules applying to the merger and acquisition (M&A) of Polish automobile firms; and (ii) the harmonization of product standards – for example, if passenger cars and commercial vehicles were tested in Germany and France, that should also be acceptable to other EU countries. One person with experience at the headquarters of the Daewoo Group in Europe was made responsible for dealing with these issues. He co-operated with almost all of the European market co-ordinators as a specialist on Korea. In developing a proposal, the director co-operated with experts who possessed the relevant technical and market knowledge. The director's political task was: (i) creating and maintaining relationships with the EU's political units – that is, contact with politicians, the treatment of political issues, and the supervision of political decisions concerning the European automobile market in countries such as Poland and the Czech Republic; and (ii) mobilizing resources in DMC when a political proposal or decision was being dealt with.

He divided the activities into internal (within DMC) and external ones. The director studied official journals, financial releases and reports on the European automobile industry. This task took up more than 15 per cent of the director's time. Part of his remaining time was devoted to studying protocols, media, competition and consumers. One fundamental area mentioned by the director dealt with building and maintaining social relationships. These duties had a social nature – they covered areas such as engagement in formal social arrangements and meetings specifically arranged for people from Poland and other EU countries. The director explained that those contacts were mainly activated when a political issue was at stake. The time spent in social and official meetings was 50 per cent, with 10–15 per cent being devoted to telephone contact. The crucial factor lay in his and his colleagues' technological and market knowledge on issues such as the customers' needs, the competitive rules, customers' duties, and new political suggestions and decisions. The acceptance of a proposal for discussion in a committee was related to the matter of how a proposal is adapted to the EU's political values. If the EU or the media does not act on a political question, the managers should collect enough technical and market information to convince the European Commission (EC) that there is a need for a political decision in the interest of consumers.

The aim of DMC was the deregulation of private handling of production plants. DMC aimed at merging freely with other automobile manufactures. In East European countries, for example, the manufacturers could choose from either M&A or joint venture alternatives. The DMC argument was based on the economic effects of such deregulation. The decision could reduce the costs for the automobile industry and subsequently reduce car prices and increase the quality of service to customers. The initiative was taken by Chairman Kim of the Daewoo Group in 1994–5, long before the issue was discussed in the media or by any political institution. The DMC director formed an *ad hoc* group to put forward a suggestion to the Finance Commission and the Industrial Policy Commission of Parliament. The committee involved in the question agreed and made the proposal to the Commission, since the proposal was in line with consumer welfare policy. The proposal was sent to Parliament but there, the different political parties and some other business groups resisted, because they were against deregulation in general. Interest groups, such as automobile makers – namely, Fiat, Renault, Volkswagen, Ford and General Motors (GM) – were also against it because this type of deregulation would give more freedom to the newcomers.

The socio-political actions of DMC are displayed in Figure 7.2. The figure is based on the subjective evaluation of DMC's commitment and trust-building

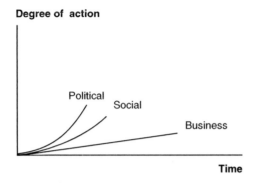

Figure 7.2   Commitment and trust-building actions in the pre-establishment phase

actions to gain legitimacy. These concern the three main areas of business, and social and political networking. DMC participated in a high degree of bidding competition to merge two Polish automobile companies. The competitor GM suggested a plan to the committee to expand the modern manufacturing plants in combination with a 33 per cent reduction in the number of employees at FSO and FSL. DMC suggested an advanced plan to expand the modern manu-facturing plants without decreasing the number of employees. On the other hand, Chairman Kim and a director of a sales subsidiary, Byung-Il Chun, attempted to negotiate with leaders of three trade unions (the labour union, metals union and engineering union) to convince them of its guarantee to maintain the current number of 20,000 employees during the first three years of its establishment. In return, they asked for a no-strike guarantee in the manufacturing plants for a period of five years.

DMC had wide experience of negotiating with the government to support their employment policy to maintain a particular number of employees during the period of economic development in Korea. Chairman Kim negotiated frequently with the Polish government to set up a duty-free system for DMC during the first two years of its establishment. The Polish taxation system still had a 35 per cent tax on imported foreign products. As a result, both the Polish government and the trade unions accepted Chairman Kim's suggestions. DMC won the battle and established three subsidiaries in Poland. From the perspective of DMC, the company had invested a minimum amount of capital in these subsidiaries to maintain the previous manufacturing and marketing channels over a short period. This result was surprising news for the Polish people and the company's Western competitors. It led to the development of DMC's posi-tion in the Polish market. Consequently, DMC used its political knowledge to win the competition with GM. This impelled DMC to accelerate its establish-ment of two manufacturing subsidiaries and a sales subsidiary in order to demonstrate a strong commitment in Poland.

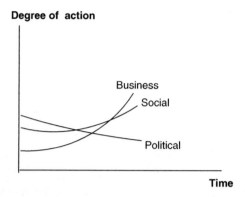

Figure 7.3 Commitment and trust-building activities in the establishment phase

Figure 7.3 demonstrates the commitment and trust-building activities for the establishment phase. As can be seen from the figure compared to Figure 7.2, there has been a change in both the business and socio-political actions. The case presentation above contains facts manifesting an increase in industrial commitments, and there are also other changes. Political actors are active, but to a lesser degree. The social relationship with highly-positioned political actors has decreased.

In June 1995, the first Polish company, FSL, was merged, and the company renamed Daewoo Motor Poland (DMP). The company manufactured a few commercial vehicles – namely, minibuses and trucks. DMC had a 71.4 per cent share of ownership, and the remainder (28.6 per cent) was owned by the Polish government. In November 1995, the second Polish company, FSO, was also merged with DMC, which was then called Daewoo-FSO. DMC had a 68.4 per cent share of the ownership, and the remaining 31.6 per cent was also owned by the Polish government. DMC had established a certain position in the market through M&A and attempted to reorganize two companies with manufacturing plants. Daewoo-FSO attempted to establish a spin-off of twenty-two joint-venture (JV) firms with Korean component firms in order to become a self-financing independent firm. The established JV firms could be more technically orientated component suppliers than the previous suppliers to FSO. Daewoo-FSO had 51 per cent stock share, and the Korean partners held 49 per cent in the twenty-two JV firms in Poland. Daewoo-FSO continued to develop relationships with sixty-one component suppliers – 23 Polish suppliers, 15 firms related to Fiat component suppliers, and 23 other suppliers.

The development phase

The CEO of Daewoo-FSO and the Daewoo Group in Poland, Jin-Chul Suk, directors of Daewoo-FSO and the Daewoo financial company attempted frequently to contact the Ministry of Finance in order to establish a new monthly payment system. They made efforts to contact government officials and the Financial Committee of Parliament to stress that market promotion (rather than paying cash) for new cars would have a positive effect on increasing economic development. Under the communist economic system, people had to pay 100 per cent cash when they wanted to buy a new car. The new payment system involved paying 50 per cent of the car price in cash and the remaining 50 per cent via a monthly payment. Subsequently, Centrum Daewoo sold a great number of cars – 137,383 in 1997. In particular, Daewoo-FSO increased its sales of different models of passenger cars – namely, Polonez, Tico, Larnos and Rexia, to both private and state-owned organizations. DMP sold a large number of commercial vehicles to fire stations, the police, government agencies and private manufacturers. Daewoo-FSO and DMP had 27.8 per cent share of passenger cars and 78 per cent share of commercial vehicles in the Polish market in 1999. The production of Daewoo-FSO in this period had rocketed from 98,416 units in 1996 to 197,226 units in 1999. Even DMP increased its sales, from 14,171 units in 1996 to 37,484 units in 1997.

Figure 7.4 illustrates the development of trust-building and commitment activities for the development phase, and shows how political actions have decreased compared with the pre-establishmet phase. But it remains at a stable level as there are problems in enforcing DMC's interaction with political organizations. The social commitment, on the other hand, has not decreased in a similar manner. Naturally, the industrial commitment is increasing as the size of investment increases. However, the increasing industrial commitment in

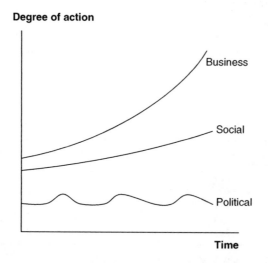

Figure 7.4 Commitment and trust-building activities in the development phase

these phases contained interactions with different types of social and political actors. Being more specific, the following sections are devoted to discussions about the content of the relationship between the DMC and these actors.

The process of interaction with leaders of trade unions

The strong position of the trade unions forced the DMC to discuss different pertinent issues with the leaders of trade unions. Issues under discussion were, for example, the symbolic flag of Daewoo-FSO, the modernization plan of Daewoo-FSO, and the detailed plan for constituting the management. Management covered issues such as a rational combination of Korean and Polish employees, and personnel situations in the organization structure. From the Korean perspective, DMC had a more strategic approach than FSO to achieving a high degree of respect from the members of trade unions and from Polish consumers. Daewoo-FSO preserved tradition and business ideology. Polish people perceived Daewoo-FSO to be a local Polish company. (See Table 7.1.)

In 1998, a problem arose from within the labour movement. One of the reasons was that the newly employed Korean chief executive officer (CEO) of the component firm JV planned to reduce the number of employees from 389 to 251. This was to follow a rationalization programme after the expiry of the contract period between Daewoo-FSO and their trade unions, during which the current number of employees had been maintained. But the members of the trade unions demonstrated strongly against both the Korean CEO of JV and Daewoo-FSO itself. The protest from the labour movement began to spread seriously throughout almost all the JV firms. Subsequently, the employees decreased car production. They even stressed that the firms did not have full schedules for manufacturing components for the following month. The CEO of Daewoo-FSO, Jin-Chul Suk, and a few directors immediately consulted with the leaders of trade unions and the Ministry of Labour. The aim was to solve the problems with the labour movements. As a result, the CEO of the JV managers resigned and the problems were resolved.

Table 7.1   Investment situation and DMC's plans in Poland (in USD millions)

Company/Period	1995–7	1998	planned total Investment of sum (–2001)
Daewoo-FSO	829.8	336.2	1,523
(Contract)	(398.5)	(285.5)	(1,121)
DMP Lublin	101	148	501
Andoria	6.2	20.7	245.5
Total	937	504.9	2,269.5

Note: The component firm JV was included in the sum of total investment.
Source: Daewoo-FSO.

The relationship with social-cultural and aid organizations

In 1995–7, DMC invested USD 829.8 million in Daewoo-FSO, USD 101 million in DMP and USD 6.2 million in Andoria (a diesel engine manufacturing company) – thus the company invested a total of USD 937 million in the manufacturing subsidiaries in Poland over that period. The size of the investment was more than twice that of the total amount of the contractual investment planned with the Polish government. The investment in twenty-two JV component firms was included in the sum of total investment. DMC was to invest a total of USD 2.27 billion in their Polish subsidiaries by 2001 to develop their business. Investment plans were focused on developing a central position for the European business headquarters, which would be established by the Polish subsidiaries of DMC as a base for implementing the global strategy of the Daewoo Group. These were important for the trade unions as they increased: (i) employment and the total number of members; (ii) the income of members and therefore also the union; and (iii) their power in both Poland and the EU.

In 1997, DMC gained significant marketing support for its operation by sending seventy-two medical personnel (doctors and nurses) from Aju University Hospital to DMP in Lublin. The medical service teams were based at DMP and performed medical services for people living in different districts of Lublin. DMC wanted to show that it was not only looking for short-term business profit. The firm wanted to strengthen its legitimacy by convincing the Polish people that it would take care of people's health. This might have led to a greater potential for Polish people to become familiar with Daewoo Motor's brand name in Poland. Moreover, DMC arranged various kinds of cultural activities – for example, a celebration of Children's Day on 5 May and an annual Korean fashion show. Children's Day included a performance of several shows and games in which both Korean and Polish families' children participated. The Korean fashion show displayed various kinds of Korean traditional and modern clothes, and was performed by both Korean and Polish fashion models. The celebration of Children's Day and other cultural activities – for example, music concerts and cultural festivals, were shown on Polish TV. The activities were organized with co-operation between DMC, the Korean–Polish culture association and other organizations.

Daewoo-FSO activities that demonstrate its social responsibility, in terms of employment or technology and foresight, have strengthened DMC's market position. These investments facilitated the firm's contact and co-operation with the political and social-cultural units in Poland. According to the director, relationships with these organizations and social contacts with politicians are necessary. The socio-political interactions are also directly connected with the firm's industrial commitment to society. The firm's specific market knowledge, its products and technology have a direct effect on socio-political relationships.

One major task of the directors is to identify and explain the socio-political benefits of proposals. In the case of 'Successful establishment of Daewoo-FSO', for example, the technical solutions, along with the political and socio-economical applications of the proposal, ensured that DMC succeeded in receiving permission to increase production capacity and ISO 9002 from EU Automotive Manufacturing Association (AMA). The managing director explained that it was because of these activities and the position of Chairman Kim of the Daewoo Group and Daewoo-FSO's CEO in EU's political organization that the firm received such permission.

Discussion

The market in this chapter is presumed to consist of the two business and socio-political market arenas. The view is constructed on the three elements of legitimacy, commitment and trust. These three elements are employed as analytical tools for the understanding of the internationalization process for MNCs' socio-political behaviour. Following this theoretical construction, the case study exposed how the Korean firms acted in Poland. In reaching the goal, as the case manifested, DMC behaved in a specific way.

DMC undertook socio-political activities to gain specific political support. The reasons why the firms proceeded in a certain manner, as explained in the case, depended on their commitment and knowledge. The results of the socio-political action process will be presented in more detail in Table 7.2. As illustrated in Table 7.2, DMC recognized socio-political activities as a critical part of their market entry and interaction. The Daewoo Group and DMC had several years of experience in socio-political matters. All three subsidiaries made a large organizational commitment – a commitment that extended not only to Brussels, but also to Paris, Munich and Warsaw, to interact with the EU AMA. DMC, for example, had organized groups and individuals in several foreign countries in order to influence politicians there. The case illustrated how political units, in their relationships with EU politicians, mobilized both internal and external resources to affect the political decisions in the matter of privatizing, modernizing and harmonizing the rules for M&A and the establishment of JV firms. The investment in a unit in Brussels and in other EU countries was claimed to be necessary in order to get closer to the socio-political actors.

One crucial factor in the interaction with socio-political actors was related to the firms' internationalization; that is, the level of business internationalization they have reached and their international market involvement. DMC made large direct investments in several EU countries. The case illustrates that the more progress a firm has made in the internationalization of its business activities, the more advanced is its socio-political commitment. Besides this, there was

Table 7.2 A summary of DMC's socio-political activities

Types of activities/ major questions in the relationships	Dyadic relationship	Development of business networks	Impact on the focal business relationship
Individual and organizational activities for political action	• Contacts and negotiations with the Ministers, Commission, Committees and Parliament in the matter of modernizing, privatizing and harmonizing the rules for M&A and the establishment of JV firms • Creation of a duty-free system during the first two years of establishment • A new financial institution setting out the monthly payment system • Daewoo-FSO products became the government's official cars	• The establishment of DMP, Daewoo-FSO and Centrum Daewoo • The development of 40 new dealer relationships and interconnecting with FSO's 180 dealer relationships • DMC invested USD 937 million in their subsidiaries and established 22 JV firms in 1995–7 • Daewoo-FSO developed relationships with 61 other component suppliers • Building strong relationships with the government and 585 dealers and after-rules centres	• The major activity is to have an influence on establishing a certain position in the market and developing customer relationships • Production activities start with marketing, for a high level of customer service • Customers were convinced by the rapid development of Daewoo-FSO and were interested in buying modern passenger cars • A monthly payment system created a very positive effect in marketing and promoting the goods
Interaction with the leaders of trade unions	• Frequent contact and negotiations with the leaders of trade unions to convince them of its guarantee to maintain the current number of 20,000 employees, and no strikes for five years • Collaboration with the leaders in designing the new flag of Daewoo-FSO	• DMC designed a flag for Daewoo-FSO in collaboration with the leaders of trade unions, to develop business operations • Daewoo-FSO organized 12 divisions in combination with the management team constituted by both Korean and Polish directors	• The customers perceived a high degree of trust in connection with Daewoo-FSO's business and service • Daewoo-FSO manufactured a modernized model of the Polonez as an improved model for customers

Table 7.2 (Continued)

Types of activities/ major questions in the relationships	Dyadic relationship	Development of business networks	Impact on the focal business relationship
	• Making a decision to harmonize personnel management	• 1,300 Polish employees participated in OJT (on-the-job-training) in Korea	• Low psychic distance between intermediary dealers and customers • To show harmony in socio-political agreement and commitments
Interaction with the EU Automotive Manufacturing Association (AMA)	• Three people in Brussels, Paris, Munich and Warsaw • The major activity is how to adapt production activities to the rules of EU • Difficult to find out how specific production can support their business	• The development of modified production in combination with DMC's European production networks to support their business • Daewoo-FSO established an R&D unit in Poland and interconnected with five international R&D units	• Daewoo-FSO obtained ISO 9002 from the EU AMA to impress customers with a high technological standard • Daewoo-FSO gained permission for a production capacity of 500,000 cars per year from the EU AMA to show business competence
Interaction with socio-cultural and aid organizations	• Aju University Hospital sent 72 medical staff to DMP in order to perform medical services in Lublin • Acquisition of a professional football team, Legia • Organizing and arranging a children's day, fashion show and cultural festivals	• Socio-cultural interaction with the Polish people • Interaction with other socio-cultural units • Extensive knowledge of how the activities of socio-cultural aid will affect the firm's business operations	• 820,000 membership cards were distributed to customers and 1,914 people participated in the marketing campaign 'a new life begins with a Reganza car', to increase socio-cultural investment.

also a relationship between the firm's business and socio-political commitments. The high business commitment of the three DMC subsidiaries became an idiosyncratic investment leading to, for example, an increase in reputation, which affected their socio-political relationships. As the DMC's three subsidiaries show, a higher level of business commitment induced the gain of a more idiosyncratic effect in socio-political interaction. The firm's social activities towards Polish society and trade unions had an effect on the trust and legitimacy of the firms and finally caused an increase in sales.

Conclusions

The fundamental conclusion of this study is that socio-political actors always have an impact on businesses and therefore the study of the business market implies the inclusion of these actors within the boundary of the firm's activities.

While the level of involvement in business activities explains the process of commitment in the business market, the commitment in the socio-political market is identified as a variable one, ranging from adaptable to influential (Hadjikhani and Ghauri, 2001). This study also pinpoints a firm's socio-political competency, which directs the firm's management activities. In this vein, the firm's competency and managerial behaviour can apparently be divided into business and socio-political areas (see also Welch and Wilkinson, 2004). These socio-political competencies are an essential part supplementing the business activities. As an outcome of idiosyncratic relationships, the commitments in both areas diffuse and affect the relationships in each area. The higher the 'pure' business commitment, the greater the firm's legitimacy and trust among socio-political actors and people in foreign markets, and the more influence the firm will gain. For its competency, the higher the influence, the more privileged the firm is in its business activities.

The above conclusions reveal another fact. While some studies refer to the homogeneous (Miller, 1992) or unidirectional impacts of governments (Kogut, 1991) and others concern the economic model for the dyadic view (Boddewyn, 1988; Jacobson *et al.*, 1993; Ring *et al.*, 1990), this study provides a different dimension. Following the role of specificity in the network, the variation in legitimacy behaviour is explained by the differences in the firms' commitment, trust and experiential knowledge. The act of exerting influence, either as a result of socio-political actors' actions or the acquisition of certain people is explained as being connected to the aspect of specificity in the relationships. The more political competency (through knowledge and commitment) an enterprise has, the more influence it gains and the more specific support it can receive. On the other hand, a low level of political knowledge and commitment leads to low levels of influence and the enterprise will perceive the relationship as coercive.

Distinguishing political from business networks leads to another conclusion that concerns an aspect of the socio-political agenda. In gaining specific support from socio-political actors, the management function of the business actors is to generate a proposal that can convince political actors. The uniqueness of this agenda can be explained by the management's competency to incorporate the business resources into the needs of the socio-political actors. As stated earlier, the network defined for this study was a set of actors – enterprises, media, governments, specialists and public opinion – linked to one another in various exchange relationships (see also Hadjikhani, 2000, 1996; Welch and Wilkinson, 2004). The interaction of these elements is explained as creating a loose network. Contrary to an industrial network (Ford, 1990), the business network in this study is contingent upon a situation in which the position of political actors is not stable, where the values of the connected actors are changing. Managers devote their time to finding out about the political actors' positions, about newcomers and new ideas released in the media, and even about their competitors' political activities. These commitments are in place to elaborate new relationships and keep the old ones alive.

References

Andersen, S. and K. A. Eliassen (1996) *The European Union: How Democratic Is It?* (London: Sage).

Barros, P. P. and T. Nilssen (1999) 'Industrial Policy and Firm Heterogeneity', *Scandinavian Journal of Economics*, 101(4), 597–616.

Boddewyn, J. J. (1988), 'Political Aspects of MNE Theory', *Journal of International Business Studies*, 14(3), 341–62.

Bolton, G. A. (1991) 'Comparative Model of Bargaining: Theory and Evidence', *American Economic Review*, 81, 1096–136.

Chaudri, V. and D. Samson (2000) 'Business–Government Relations in Australia: Cooperating Through Task Forces', *Academy of Management Executives*, 14(3), 19–29.

Conner, K. R. (1991) 'A Historical Comparison of Resource-based Theory and Five Schools of Thoughts within Industrial Organization Economics', *Journal of Management*, 17, 121–54.

Crane, A. and J. Desmond (2002) 'Social Marketing and Morality', *European Journal of Marketing*, 36(5/6), 548–69.

Crawford, V. (1982) 'A Theory of Disagreement in Bargaining', *Econometrica*, 50, 607–37.

Denekamp, J. (1995) 'Intangible Assets, Internationalization and Foreign Direct Investment in Manufacturing', *Journal of International Business Studies*, 26(3), 493–504.

Dixon, D. F. (1992) 'Consumer Sovereignty, Democracy and the Marketing Concept', *Canadian Journal of Administrative Science*, 9(2), 116–25.

Egelhoff, W. G. (1988) 'Strategy and Structure in Multinational Corporations: A Revision of the Stopford and Wells Model', *Strategic Management Journal*, 9, 1–14.

Esping-Andersen, G. (1985) *Politics against Markets, The Social Democratic Road to Power*, (Princeton, NJ: Princeton University Press).

Fligstein, N. (1990) *The Transformation of Corporate Control* (Cambridge, Mass.: Harvard University Press).

Ford, D. (ed.) (1990) *Understanding Business Markets: Interaction, Relationships, Networks*, (London: Academic Press).

Ghauri, P. N. and P. R. Cateora (2006) *International Marketing* (2nd edn) (London: McGraw-Hill).

Ghauri, P. N. and Holstius, K. (1996) 'The Role of Matching in the Foreign Market Entry Process in the Baltic States', *European Journal of Marketing*, 30(2), 75–88.

Hadjikhani, A. (1996) 'Sleeping Relationship and Discontinuity in Project Marketing', *International Business Review*, 3(5), 319–37.

Hadjikhani, A. (2000) 'The Political Behavior of Business Actors', *International Studies of Management and Organization*, 30(1), 95–119.

Hadjikhani, A. and P. Ghauri (2001) 'The Internationalization of the Firm and Political Environment in the European Union', *Journal of Business Research*, 52(3), 263–75.

Hanf, K. and T. A. J. Toonen (eds) (1985) *Policy Making in Federal and Unitary Systems*, (Dordrecht: Kluwer).

Jacobson, C. K., S. A. Lenway and P. S. Ring (1993) 'The Political Embeddedness of Private Economic Transactions', *Journal of Management Studies*, 30(3), 453–78.

Johanson, J. and J.-E. Vahlne (1990) 'The Mechanism of Internationalisation', *International Marketing Review*, 7(4), 11–24.

Keillor, B. D. and T. M. Hult (2004) 'Predictors of Firm-Level Political Behavior in the Global Business Environment', *International Business Review*, 13(3), 309–29.

Keillor, B. D., G. W. Boller and O. C. Ferrel (1997) 'Firm-Level Political Behaviour in the Global Market Place', *Journal of Business Research*, 40, 113–26.

Kogut, B. (1991) 'Country Capabilities and the Permeability of Borders', *Strategic Management Journal*, Summer Special Issue, 12, 33–48.

Korbin, S. (1982) *Managing Political Risk Assessment: Strategic Responses to Environmental Changes* (Berkeley, Calif.: University of California Press).

Lee, J.-W. (1991) 'Swedish Firms Entering the Korean Market – Position Development in Distant Industrial Networks', Dissertation, Department of Business Studies, Uppsala University.

Maddison, A. (1991) *Dynamic Forces in Capitalist Development, A Long-run Comparative View* (Oxford University Press).

Miller, K. D. (1992) 'Industry and Country Effects on Managers' Perspective of Environmental Uncertainties', *Journal of International Business Studies*, 24(1), 693–714.

Nowtotny, K., D. B. Smith and H. M. Trebling (eds) (1989) *Public Utility Regulation: The Economic and Social Control of Industry* (Boston, Mass.: Kluwer).

O'Shaughnessy, N. (1996) 'Social Propaganda and Social Marketing: A Critical Difference', *European Journal of Marketing*, 10/11, 54–67.

Potters, J. (1992) *Lobbying and Pressure* (Amsterdam: Tinbergen Institute).

Ramaswamy, K. and W. Renforth (1996) 'Market Intensity and Technical Efficiency in Public Sector Firms: Evidence from India', *International Journal of Public Sector Management*, 9(3), 4–17.

Ring, P. S., S. A. Lenway and M. Govekar (1990) 'Management of the Political Imperative in International Business', *Strategic Management Journal*, 11, 141–51.

Rose-Ackerman, S. (1978) *Corruption* (New York: Academic Press).

Scott, J. (1994) *Social Network Analysis, A Handbook* (London: Sage).

Streeck, W. (1992) *Social Institutions and Economic Performance: Studies of Industrial Relations in Advanced Capital Economics* (London: Sage).

Welch C. and I. Wilkinson (2004) The Political Embeddedness of the International Business Network, *International Marketing Review*, 21(2), 216–31.

8

Internationalization Strategies Realized by Incumbent Firms as an Industry Evolves into a Global Oligopoly: The Case of the Pharmaceutical Industry

Amanda Jane Langley, Nada Korac Kakabadse and Stephen Swailes

Introduction

This chapter has been driven by a desire to understand how the internationalization strategies of firms that arrived at different strategic outcomes evolved as an industry transformed itself into a global oligopoly. Evolutionary theory was applied to answer the question 'How did strategic actions regarding the internationalization of incumbent firms evolve as the pharmaceutical industry transformed itself into a global oligopoly?' The pharmaceutical industry was chosen as the focus of study because of changes in its structure and the pressure that firms are facing because of increasing health care reforms.

Data was collected from published documentary sources and analysed qualitatively to explore how internationalization strategic actions evolved for six medium-sized pharmaceutical firms. The chapter shows that all firms in the sample realized strategic internationalization actions and maps how these evolved chronologically between 1992 and 2002. Despite the firms in the sample being medium-sized European pharmaceutical firms, and with three of them – Shire, Galen and Bioglan – sharing very similar characteristics, each has evolved unique patterns of internationalization strategy development and arrived at different strategic outcomes. The chapter shows that, during periods of industry anxiety, firms realize internationalization strategies in different ways in their attempt to survive.

The move towards the creation of global oligopolies is happening in many industries (Galambos, 2001) despite the risks associated with internationliza-tion. Operating in new countries risks potential problems relating to commercial, cross-cultural, currency/financial, and country (political and legal) risks (Cavusgil, 2006). The process of industries becoming global oligopolies is driven by competitive anxiety at the firm level, as internationalizing firms seek to compete and survive in rapidly changing environments.

Literature review

Global trends can shape the evolution of firms (Lawrence, 2002). The emergence of new technology and regulatory changes, together with increased merger activity and other internationalization strategies, leads to firms consolidating and globalizing in order to survive and to maximize profits. Industry globalization results from two main causes – industry concentration and internationalization – which have partially been driven by increasing financial concentration (Chesnais, 1993). Other factors that have contributed to industry globalization include the globalization of regulation, the evolution of multinational corporations, emerging technologies, cross-border strategic alliances, and new forms of global communication such as the World Wide Web (Braithwaite and Drahos, 2000; Chesnais, 1993; Tolentino, 2000).

According to the structure–conduct–performance (S–C–P) paradigm, changes in industry structure influence strategy evolution in incumbent firms (Bain, 1956; Mason, 1959) and that those strategies shape industry structure (Scherer, 1980). This suggests that firm strategies lead to industry consolidation, which leads to changes in incumbent firm strategies; and so the cycle continues. Firms in a fragmented industry can potentially make strategic choices that move the industry towards consolidation and thus increased concentration (Porter, 1980). One of the ways in which firms in a fragmented industry can work to increase market power is through merger and acquisition activity within an industry (Porter, 1980). Other globalization strategies include licensing, the establishment of R&D facilities and exporting as firms seek to enter non-national countries (Porter, 1980; Taggart, 1993).

The pharmaceutical industry has evolved into a global oligopoly. In the 1980s Jones and Cockerill (1984) described the pharmaceutical industry as being highly fragmented. One of the reasons for this fragmentation was that, because of the inherently volatile nature of the research and development (R&D) process, few firms would contemplate taking the additional risks associated with merger and acquisition (M&A) activities (Taggart, 1993). However, consolidation of the larger pharmaceutical firms started in the late 1980s, with the formation of organizations such as SmithKline Beecham and Bristol Myers Squibb, and increased in pace during the 1990s (Heracleous and Murray, 2001; Matraves, 1999; Pursche, 1996). This was followed by a number of 'megamergers' which resulted in the formation of firms such as GlaxoSmithKline and Novartis (Matraves, 1999; Schmidt and Ruhli, 2002) leading to 'consolidation of firms at the top' (Matraves, 1999, p. 188).

Galambos (2001:19) appears to indicate that the consolidation of large firms is the phenomenon that has created a pharmaceutical industry which demonstrates the characteristics of an 'oligopoly in various therapeutic categories', and what appears to be a transition towards the creation of a global oligopoly (Kettler, 2001). Strategies including international co-operative agreements and foreign direct investment (FDI) have led to increasing global flows of finance, which has been reinforced by the globalization process (Walsh and Galimberti, 1993). The result has been that the leading pharmaceutical firms operate either internationally or globally (Matraves, 1999).

Regulation of the pharmaceutical Industry

Cavusgil (2006) identified legal issues as one of the potential risks that firms encounter when entering new countries. For pharmaceutical firms this is accompanied by the strong regulation of the industry; the pharmaceutical industry is possibly more regulated than any other industry (Earl-Slater, 1993). Alongside the need to be innovative, pharmaceutical firms have also been put under pressure as a result of health care reforms. These have included increasing costs of clinical trials and measures to reduce health care expenditure (Boscheck, 1996). Pressure for changes in regulatory regimes were increasingly the result of action taken at the European level, starting with the 1965 EC Directive on Medicinal Products. This directive focused on issues relating to safety, quality and efficacy (Braithwaite and Drahos, 2000). The European influence also affected decisions relating to pricing and reimbursement, and has initiated a drive towards European rather than national approval of new drugs reaching safety standards (Kanavos and Mossialos, 1999).

Regulation in some countries with regard to issues of cost and reimbursement affects how a product is supplied to the market (Kanavos and Mossialos, 1999), and has been combined with issues relating to shortening product life-cycles. In order to overcome the resulting financial pressures, pharmaceutical firms have sought to increase profits through the development of new products, and through geographical expansion (Matraves, 1999; Taggart, 1993; Walsh and Galimberti, 1993).

Objectives of the study

This chapter draws on evolutionary theory in order to understand the strategy processes relating to the international development of pharmaceutical firms during recent changes in the industry's structure. Evolutionary theory underpins the concepts of incremental and emergent strategies (Lynch, 1997; Mintzberg *et al.*, 1998). Although there have been a variety of studies of pharmaceutical strategies, there appears to have been a limited examination of strategy evolution or temporal patterns of internationalization strategy development with regard to the pharmaceutical industry's medium-sized pharmaceutical firms.

This leads to the first research objective:

Objective 1 To identify the internationalization strategic actions realized by medium-sized pharmaceutical firms as the pharmaceutical industry evolved into a global oligopoly

There has been some recent research into strategy formation and processes in large pharmaceutical firms (Schmidt and Ruhli, 2002) and studies of US market entry strategies by pharmaceutical firms (Bogner *et al.*, 1996; Howell, 2004). Rugman and Brain (2004) explored the regional strategies of pharmaceutical firms, but the emphasis was on R&D. There is therefore a gap with regard to how pharmaceutical strategies, not just those focused on R&D, evolve internationally.

This leads to the second research objective:

Objective 2 To explore patterns of evolution of strategic action in different geographical markets for a set of medium-sized pharmaceutical firms as the pharmaceutical industry evolved into a global oligopoly

Although each geographical market may present different regulatory hurdles, the R&D technology itself is not difficult to transfer (Matraves, 1999). This can be achieved through various strategies including the establishment of wholly-owned marketing and distribution networks, entering into co-operative arrangements such as joint ventures and co-marketing agreements, outsourcing of activities, or the establishment of overseas R&D facilities (Walsh and Galimberti, 1993; Matraves, 1999; Schmidt and Ruhli, 2002).

This leads to the third research objective:

Objective 3 To compare overall patterns of internationalization strategy evolution for a set of middle sized pharmaceutical firms.

Methodology

A qualitative methodological approach has been adopted for this research, based on a categorization of grand strategies and strategic actions realized by firms in the pharmaceutical industry (Langley *et al.*, 2005). This categorization was adapted to identify the internationalization strategic actions realized by firms.

In this chapter, internationalization strategic actions are those that were identified during the coding of the empirical data in relation to grand strategies and refers to entry into and development within international markets. The data collection focused on strategic actions relating to merger and acquisition (M&A), network and acquisition based product development (NABPD), joint venture (JV), organic concentric diversification (OCD) and organic growth (OG).

The timescale for data collection was 1992–2002, a period that witnessed increased concentration and globalization in the pharmaceutical industry. The sources of empirical data were *Scrip*, a major pharmaceutical industry trade journal, the *Financial Times* and the *Mergerstat* database.

Before undertaking the analysis to identify the strategic internationalization actions realized by each firm, N5 was used to retrieve all the articles relating to each firm being analysed. The text was then analysed using the coding instrument as shown in Table 8.1. The geographical location was recorded for each of the realized strategic actions. The relevant data from the coded articles were then recorded on spreadsheets in the format shown in Table 8.2.

Coding of the data identified four geographical themes with regard to internationalization/globalization markets. These were Western Europe, Central and Eastern Europe (CEE), the USA and 'the rest of the world'. From the tabulated results, summaries of the internationalization strategic actions for each firm are provided.

Pearce (1982) proposed that actions constituting a grand strategy should happen over a minimum period of five years. Therefore strategic actions are considered to form a strategy for the relevant geographical market if they are realized for at least three years in any five-year period. The definition of strategic actions that constitute a grand strategy still leaves the issue of strategic actions that were realized but were not necessarily part of an overall grand strategy; that is, there was no consistency in action (Kay, 1993; Mintzberg *et al.*, 1998). These have been classed as incremental strategic actions (Lindblom, 1959, 1979; Quinn, 1980, 1991); that is, those strategic actions that were not realized with the frequency required to be a strategy.

A limitation with the data collection is a lack of historical data. This limitation was accepted in Webb and Pettigrew's (1999) paper into temporal patterns in strategy development. They acknowledged in their study that the timescale for data collection means that the researcher can only refer to actions that occurred during the period under study.

Population selection

Purposive non-probability sampling was used to select a small number of cases (Patton, 2002). During initial interviews to familiarize themselves with the pharmaceutical industry, the authors found that pharmaceutical organizations that were growing to medium size were usually being bought by larger firms or failing to survive. Therefore, the cases selected through the purposive sampling were those considered to be 'information rich' (Patton, 2002, p. 46) in that they represent medium-sized pharmaceutical firms that arrived at different strategic outcomes.

Table 8.1   The coding instrument

Grand strategy	Definition of grand strategy	Criteria for exclusions/ qualifications
Network and acquisition based product development (NABPD)	Refinement of an existing pharmaceutical-related product or development of a new product or licensing of a product through a co-operative arrangement. This also includes the acquisition of a single product or number of products	Co-operative arrangements include strategic alliances, licensing-in agreements, co-promotion, co-marketing, franchising, consortia and outsourcing This category does not include co-operative arrangements relating to the external raising of finance. that is, those that involve the firm having equity placed in it by another firm
Organic concentric diversification (OCD)	Internal generation of a separate business	The spin-off or creation of a new business, which must be solely owned
Organic growth (OG)	Corporate expansion activities which include an increase in assets and expenditure	This does not include the acquisition or merger of businesses or increases that are product specific
Merger and acquisition (M&A)	This incorporates horizontal integration, vertical integration and M&A concentric diversification strategies	This is focused on the merger or partial or full acquisition of a business. The text must be interpreted in light of the firm's main area of activity
Joint venture (JV)	The creation of a third 'daughter' firm by two partner firms.	Text will refer to either joint venture or jointly owned affiliate company, business, business unit or spin-off

Sources: Based on Pearce II and Robinson (1994); Langley *et al.* (2005)

The sampling frame was the top 200 pharmaceutical firms by turnover at the end of 2000, as identified in the *Scrip Pharmaceutical League Tables* (PJB Publishing, 2001). The firms were tracked to identify their strategic outcome at the end of the period being studied, which was 31 December 2002. Only European firms were selected, as they develop strategies in a different regulated

Table 8.2 Example of recorded data for Pierre Fabre

Year	Strategic action	JV partner	JV name	Countries
1994	Establishment of an international centre to study skin ageing	Toulouse University Hospital		France
2000	Establishment of a JV	Arriani Pharmaceuticals	Pharma Fabre SA	Greece
2000	JV production plant	Novo Nordisk	Aldapan	Tizi Ouzou region
2001	Joint research centre	CNRS – the French National Scientific Research Centre		France

Source: Compiled by authors from *Scrip* and *Financial Times*, 1992–2002.

Table 8.3 Firms in the sample

Name	Country of origin	Strategic outcome	Position in the *Scrip Pharmaceutical League Tables*, 2000
Asta Medica	Germany	Disbanded and divested	65
Pierre Fabre	France	Demerged	77
Shire	UK	Merged	80
LEK	Slovenia	Acquired	111
Galen	UK	Survived without being merged or acquired	140
Bioglan	UK	Liquidated	143

trade bloc from firms in the USA and Japan. To maximize the credibility of the analysis, a multiple case replication approach was chosen, which increased the stability of findings while allowing for generalizability across cases (Miles and Huberman, 1994). The firms in the sample are outlined in Table 8.3.

Findings: internationalization strategic actions realized by six medium-sized pharmaceutical firms during 1992–2002

Pierre Fabre

Pierre Fabre is a privately-owned French company established in 1961 (Pierre Fabre, 2003). It evolved into being 'France's second largest privately owned

Table 8.4 Summary of Pierre Fabre's strategic internationalization actions

	1992	1993	1994	1995	1996	1997	1998	1999	2000	2001	2002
M&A		✓		✓				✓		✓	
NABPD	✓		✓	✓	✓			✓	✓	✓	
JV									✓		
OCD						✓					
OG											

pharmaceutical firm' (*Scrip*, 1994, p. 13). Pierre Fabre began its strategic internationalization actions in 1992, with actions relating to both the Western European and CEE markets (see Table 8.4).

Pierre Fabre used JVs, M&A, OCD and NABPD strategic actions to extend its geographical markets. Various incremental strategic actions were subsequently realized with regard to the CEE market. With the exception of 2001, no strategic actions were reported in relation to the US market. In comparison, strategic actions relating to the Western European market were realized for the majority of the years 1992–2002. Although it realized strategic action(s) relating to 'the rest of the world' in 1995, Pierre Fabre did not start to pursue a 'rest of the world' strategy until the late 1990s. Pierre Fabre appeared to develop its most local market, Western Europe, before progressing to a wider 'rest of the world' strategy, which excluded the USA, despite it being regarded as the largest pharmaceutical sales market (Walton, 2001).

LEK

LEK is a generic pharmaceuticals firm based in Slovenia. In 2004, LEK classed itself in worldwide terms as 'a mid size pharmaceutical firm' (LEK, 2004). LEK's strategic internationalization actions started with the CEE market in 1992 followed by the US market in 1996 (see Table 8.5). None were reported with regard to Western Europe or the 'rest of the world' until 1999.

LEK focused on developing specific strategic actions in the CEE market from 1992, but until 2001 these were focused on organic growth. LEK only realized

Table 8.5 Summary of LEK's strategic internationalization actions

	1992	1993	1994	1995	1996	1997	1998	1999	2000	2001	2002
M&A										✓	
NABPD								✓		✓	
JV											
OCD					✓						
OG	✓				✓			✓			✓

strategic actions for the US market in 1996, with the establishment of a US company, and a collaboration agreement with Ethical Holdings in 1999. LEK only realized one strategic action with regard to the 'rest of the world', which was a collaboration agreement.

Asta Medica

Asta Medica was a German firm established in 1919 (Hoppenstedt Firmeninformationen, 2003) and was the pharmaceuticals division of Degussa. In 1992, Asta Medica's core business was focused on 'branded generics, original products and OTC drugs' (*Scrip*, 1992a, p. 9). Asta Medica started to realize its strategic internationalization actions in 1992 through the use of M&A and OCD to access the US market, and NABPD and JV for Western Europe (see Table 8.6).

Asta Medica did not pursue any more strategic actions for the US market until 1996, when it added to its minority share in Muro Pharmaceuticals to enable a full acquisition. This was followed in 1997 with a licensing-in agreement. In 1998, Asta Medica completed its strategic actions for the US market. The strategic actions for Western Europe evolved from 1992–2001, mainly through NABPD and JV strategic actions. The exception was in 1994, with a new plant facility in France. Asta Medica pursued only OG and NABPD incremental strategic actions for the CEE market. Asta Medica realized a strategy for the 'rest of the world' market starting in 1994 and finishing in 2001.

Shire

Established in 1986 as a biotechnology start-up, Shire evolved into a speciality pharmaceuticals firm (Guerrera and Firn, 2000). Shire started implementing strategic actions to develop itself consistently in the CEE market during the period 1994–2002 (see Table 8.7).

Shire realised a strategy for each of the four geographical markets. In 1997, the company acquired two US firms and appeared to build on its 1997 acquisition with its 1999 merger with Roberts Pharmaceutical. Shire's other strategic actions relating to the 'rest of the world' did not take place until 2001, with its merger with the Canadian firm Biochem Pharma and the 2002 investment in a global vaccine research centre, also based in Canada.

Table 8.6 Summary of Asta Medica's strategic internationalization actions

	1992	1993	1994	1995	1996	1997	1998	1999	2000	2001	2002
M&A	✓				✓	✓	✓				
NABPD	✓		✓			✓	✓			✓	
JV	✓		✓				✓		✓		
OCD	✓		✓								
OG			✓			✓					

Table 8.7   Summary of Shire's strategic internationalization actions

	1992	1993	1994	1995	1996	1997	1998	1999	2000	2001	2002
M&A						✓		✓		✓	
NABPD			✓	✓	✓	✓	✓		✓	✓	✓
JV											
OCD								✓			
OG	✓							✓			✓

Galen

The Northern Ireland integrated pharmaceutical company, Galen, was founded in 1968 (Galen Holdings plc, 2001). In 1992, Galen was reported as having a pharmaceuticals strategy focused on competing with branded generics (*Scrip*, 1992b). Galen did not realize any strategic actions outside the UK until 1997 (see Table 8.8).

From 1997 all of Galen's internationalization strategic actions were focused on the USA, primarily in the form of acquisitions. This was concentrated in the period 1998–2000, which appears to have been a very intense period for Galen's acquisition strategy.

Bioglan

The British firm, Bioglan, was founded in 1932 by Menzies Sharp as a vitamins company and fifty years later (1982) was bought by Terry Sadler. By 1989, Bioglan was operating in both generics and dermatology. The company did not realize any strategic actions relating to international markets until 1994. Prior to 1998 these were focused on Western European and US markets (see Table 8.9).

The majority of Bioglan's strategic actions for Western Europe were related to acquisitions and its NABPD strategy, together with OG and OCD strategic actions. For the US market, Bioglan combined M&A, NABPD, OCD and OG strategic actions. In 1998, Bioglan also realized strategic actions for the CEE and 'rest of world' markets. For both of these markets Bioglan only realized

Table 8.8   Summary of Galen's strategic internationalization actions

	1992	1993	1994	1995	1996	1997	1998	1999	2000	2001	2002
M&A							✓	✓	✓		
NABPD											✓
JV											
OCD											
OG						✓					

Table 8.9   Summary of Bioglan's strategic internationalization actions

	1992	1993	1994	1995	1996	1997	1998	1999	2000	2001	2002
M&A			✓	✓	✓		✓	✓	✓		
NABPD							✓	✓	✓	✓	
JV											
OCD					✓			✓			
OG								✓		✓	

strategic actions relating to a NABPD strategy. The company realised strategic actions relating to internationalization strategies for all the markets apart from those of the CEE.

Discussion

Patterns of strategic action evolution in different geographical markets

The first movers into the Western European market were Pierre Fabre, Asta Medica and Shire, who all realised relevant strategic actions in 1992 (see Table 8.10). Of the firms realizing these strategic actions, the last mover was LEK. There was a period of convergence in 2001 (five firms). The firm that diverged from this process was Galen, who did not realize any strategic actions relating to Western Europe. The reason for the convergence in 2001 may have been because of European regulatory changes during the 1990s, such as the creation the European single market and the European Medicines Evaluation Agency (EMEA) (Braithwaite and Drahos, 2000; Matraves, 1999; Taggart, 1993). These regulatory changes had the potential to encourage more strategic activity within Western Europe by European pharmaceutical firms, including the Eastern European firm LEK.

The picture for the US market is different from that of internationalization into Western Europe (see Table 8.11). Asta Medica was the first mover, in 1992, and Pierre Fabre the last mover. Apart from than, for the period 1992–5 no other firm realized strategic actions with regard to the US market apart from

Table 8.10   Strategic actions relating to Western Europe

	1992	1993	1994	1995	1996	1997	1998	1999	2000	2001	2002
Bioglan			▓	▓	▓		▓	▓	▓	▓	
Pierre Fabre	▓	▓							▓	▓	
LEK								▓	▓		
Asta Medica	▓		▓	▓	▓		▓	▓			
Shire	▓	▓	▓	▓	▓	▓	▓	▓	▓	▓	▓
Galen											

Table 8.11   Strategic actions relating to the USA

	1992	1993	1994	1995	1996	1997	1998	1999	2000	2001	2002
Bioglan					▓		▓	▓		▓	
Pierre Fabre										▓	
Galen						▓	▓	▓	▓		▓
LEK							▓	▓			
Asta Medica	▓					▓	▓				
Shire				▓		▓	▓	▓	▓		▓

Shire, who started a US strategy from 1995 that continued until 2002. From 1996, at least three firms realized relevant strategic actions in each of the subsequent years, with a period of convergence (four firms) in 1998–9.

Although there was concern in the USA in 1994 about possible health care reforms in Europe, the reality was that such reforms were already being implemented (Walton, 2001). The findings from the empirical data do not support this with regard to a reduction of internationalization strategies in Western Europe, but do appear to support it with regard to expansion into the USA, as noted by the 1998–9 period of convergence. A conclusion can be drawn that regulatory changes in Europe were shaping the strategies of European firms so that they developed a presence in the USA.

With regard to the CEE market, the first movers were Pierre Fabre, LEK and Asta Medica, with the last mover being Bioglan (see Table 8.12). There was a period of convergence in 2001 (4 firms) and a period of divergence in 1993 (0 firms). Some of the strategic actions that led to developments in the CEE market were in fact worldwide agreements that related to all the geographical markets identified in the coding. The only firms that implemented specific strategic actions were the Slovenian (CEE) firm LEK, and the German-based Asta Medica.

The first movers for the rest of the world market were Asta Medica, Pierre Fabre and Shire (see Table 8.13), though none of these realized relevant

Table 8.12   Strategic actions relating to CEE

	1992	1993	1994	1995	1996	1997	1998	1999	2000	2001	2002
Bioglan							▓				
Pierre Fabre	▓				▓		▓		▓		
Galen											
LEK	▓				▓		▓			▓	▓
Asta Medica	▓		▓							▓	
Shire			▓	▓		▓	▓		▓	▓	▓

Table 8.13   Strategic actions relating to the rest of the world

	1992	1993	1994	1995	1996	1997	1998	1999	2000	2001	2002
Bioglan							▨		▨	▨	
Pierre Fabre			▨	▨				▨	▨		
Galen											
LEK								▨			
Asta Medica			▨	▨		▨	▨		▨	▨	
Shire			▨	▨		▨	▨		▨	▨	▨

strategic actions until 1994. This compares to all the other markets, where at least one firm was involved in each from 1992. More firms were involved in the 'rest of the world' market by 1994 than those realizing strategic actions for the US market. The last mover was LEK. The only firm not realizing strategic actions relating to the CEE market was Galen. There was a period of convergence in 2000–1 (four firms).

Comparison of internationalization strategy development: geographical and temporal patterns

Shire was the only firm to realize internationalization strategies for all four markets, while Asta Medica realized internationalization strategies for all the markets apart from CEE. In comparison, LEK only realized an internationalization strategy for the CEE market, though it did realize incremental strategic actions for all the other geographical markets.

Galen only realized a strategy for one market, the USA. In comparison to LEK and Galen, Pierre Fabre did not realize internationalization strategies for either the USA or the CEE markets, but did for the 'rest of the world' and Western Europe. For the five firms compared so far, each chose a unique international strategy selection for 1992–2002. Bioglan's international strategy selection mix with regard to geographical markets is the same as that for Asta Medica, though its overall internationlization strategy evolved differently.

Asta Medica started to realize its internationalization strategy in 1992, with strategic actions relating to Western Europe, the USA and the CEE market. In comparison, Bioglan did not start until 1994, with its M&A strategic action to relating to the Western European market. With regard to national path dependency McKelvey *et al.* (2004, p. 113) said that 'institutional/country-specific factors are particularly important for explaining the different patterns visible at different levels'. This view is supported by the example of LEK, the only non-Western European firm.

LEK realized the lowest level of strategic actions with regard to Western Europe. A possible reason for this is that standards of products in CEE countries were not necessarily as high as in Western markets, but with the potential accession to the EU of countries such as Slovenia, emphasis was placed on increasing product quality to meet these standards. This may explain why LEK did not begin its strategic actions in Western Europe until 1999, and these were quite limited in comparison to the previous firms mentioned. This does not explain why the internationalization strategies of Shire, Galen and Bioglan were different as they had all originated from the UK, they were all classed as 'speciality pharmaceutical' companies (Guerrera and Pilling, 2000, p. 36) and were similar in size.

With regard to the temporal patterns of international strategy development of the sample firms, Asta Medica was a first mover for all the geographical markets, but there was no overall last mover. Specific periods of 'convergence' (firms in the sample entering the same region) can be seen for each of the geographical markets. Although there was little strategic activity with regard to the US market until 1995, 1998–9 is a convergence period, with a total of five firms in the sample realizing relevant strategic actions in this period. This was followed in 2000 by four firms realizing strategic actions for the 'rest of the world'. In 2001, four firms realized strategic actions for both the CEE and the 'rest of the world' markets.

These periods of strategic action convergence suggest a 'mimetic isomorphism' (Mintzberg *et al.*, 1998, p. 295) as the firms have become increasingly institutionalized within the environment, which led to a process of them copying the strategic actions realized by other firms. Bioglan, the firm that was liquidated, was the only firm not to be a first mover for any of the geographical markets, which suggests institutionalization with regard to Bioglan's strategic actions. In contrast, Galen, the firm that survived without being merged or acquired, was the only firm in the sample not to realize strategies for any of the geographical markets apart from the USA, therefore showing no evidence of becoming institutionalized. Asta Medica, the firm that was disbanded and divested, was the only firm to be a first mover for all the internationalization grand strategies. This therefore suggests that an emphasis on being a first mover with regard to internationalization development is not a source of competitive advantage in the pharmaceutical industry.

Although there were differences in how the strategic internationalization actions evolved, there was evidence of firms combining incremental strategic actions for developing into some geographical markets alongside specific strategies for other geographical areas (see Table 8.14). (The terms 'strategy' and 'incremental strategic actions' were defined in the methodology section).

There were some other common themes with regard to strategic choices for firms entering/developing themselves in international markets. All firms realized OCD/OG strategic actions (greenfield investments), and all realized M&A strategic actions despite the risks associated with this form of internationalization strategy. For example, entry into the US market via acquisition has been

Table 8.14   Internationalization strategies and incremental strategic actions realized

	USA	CEE	Western Europe	Rest of world
Pierre Fabre	Incremental	Incremental	Strategy	Strategy
LEK	Incremental	Strategy	Incremental	Incremental
Galen	Strategy	None	None	None
Shire	Strategy	Strategy	Strategy	Strategy
Bioglan	Strategy	Incremental	Strategy	Incremental
Asta Medica	Strategy	Incremental	Strategy	Strategy

shown to be more likely to result in exit from the market compared to entry through greenfield investments (Li, 1995). All firms realized international NABPD agreements, which appears to lend support to the network approach to internationalization (Hollensen, 2004). Despite these common themes, only two of the six firms involved themselves in JVs.

Conclusions

Evolutionary theory was applied to explore how incremental strategic actions and strategies evolved for a set of medium-sized pharmaceutical firms as the industry transformed itself into a global oligopoly. Empirical data on the sample firms' internationalization allowed us to map how internationalization strategies evolved geographically over the period 1992–2002. Despite the firms in the sample all being medium-sized European pharmaceutical firms, and with three of them–Shire, Galen and Bioglan–sharing very similar characteristics, each evolved unique patterns of internationalization strategy development and arrived at different strategic outcomes.

In relation to strategic outcomes, the findings highlighted that the firm which was a first mover for all of the geographical markets (Asta Medica) was disbanded and divested. Bioglan, the firm that developed strategies for the same geographical markets as Asta Medica, also failed to survive, with the strategic outcome of being liquidated. The firm that had an internationalization strategy for all of the geographical markets (Shire) survived in a merged format, while the firm that focused its strategy on the CEE market (LEK) was acquired. In contrast, the firm that focused on the US market (Galen) survived without being acquired or merged. In comparison to LEK and Galen, Pierre Fabre did not realize internationalization strategies for either the US or CEE markets, but did realize them for the 'rest of the world' and Western Europe. Pierre Fabre's strategic outcome was its demerger.

In addressing its objectives, this chapter was able to answer the question, 'How did the strategic internationalization actions realized by incumbent firms evolve as the pharmaceutical industry transformed itself into a global oligopoly?' It was found that all the firms in the sample realized strategies outside their country of domicile. This was achieved through strategies for specific geographical markets, with some firms combining these with incremental strategic actions (see Table 8.14). Each firm evolved its overall internationalization strategy in a unique way. With some firms, this was directed at specific international markets–for example, Galen and the USA. For others there is evidence of moving into Western Europe first before expanding to geographical markets further from the country of domicile. For example, Bioglan started in Western Europe (1994), progressed to realizing strategic actions in the USA (1996), and then two years later extended these strategic actions to the CEE and the rest of the world.

This chapter contributes towards filling a gap in the literature with regard to how pharmaceutical strategies, not just those focused on R&D, evolved internationally as the pharmaceutical industry transformed itself into a global oligopoly. By focusing on medium-sized firms that arrived at different strategic outcomes, it highlights the anxiety underpinning the strategic internationlization decisions made as they sought to survive in an industry facing challenges that included both regulatory and structural changes.

Further research might contribute towards testing the extent to which the particular patterns of internationalization strategy development identified in this study lead to certain strategic outcomes for firms operating in the global pharmaceutical industry.

Acknowledgement

Information from *Scrip* was used with the kind permission of PJB Publications.

References

Bain, J. S. (1956) *Barriers to New Competition* (London: Oxford University Press).

Baumol, W. J. (1982) 'Contestable Markets: An Uprising in the Theory of Industry Structure', *The American Economic Review*, March, 1–15.

Bogner, William C., Howard Thomas and John McGee (1996) 'A Longitudinal Study of the Competitive Positions and Entry Paths of European Firms in the US Pharmaceutical Market', *Strategic Management Journal*, 17, 85–107.

Boscheck, Ralf (1996) 'Health Care Reform and the Restructuring of the Pharmaceutical Industry', *Long Range Planning*, 29(5) (October), 629–42.

Braithwaite, John and Peter Drahos (2000) *Global Business Regulation* (Cambridge University Press).

Cavusgil, S. (2006) 'International Business in the Age of Anxiety: Company Risks', UK AIB Conference, Manchester, April.

Chesnais, F. (1993) 'Globalisation, World Oligopoly and Some of Their Implications', in M. Humbert (ed.), *The Impact of Globalisation on Europe's Firms and Industries* (London: Pinter).

Earl-Slater, A. (1993) 'Pharmaceuticals', in by P. Johnson (ed.), *European Industries Structure, Conduct and Performance* (Aldershot: Edward Elgar).

Galambos, Louis (2001) 'Global Oligopoly, Regional Authority and National Power: Crosscurrents in Pharmaceuticals Today and Tomorrow', in H. E. Kettler (ed.), *Consolidation and Competition in the Pharmaceutical Industry*, based on papers delivered at the OHE Conference, London, 16 October 2000 (London: Office of Health Economics).

Galen Holdings Plc (2001). *Galen Holdings Plc 2000 Annual Report and Accounts* (London: ICC Information Group Ltd).

Guerrera, F. and D. Firn (2000) 'Stahel Believes His Shire Horse Will Keep on Running', *Financial Times*, 12 December, 32.

Guerrera, F. and Pilling, D. (2000) 'Ugly Ducklings Swan into the Middle Ground', *Financial Times*, 7 November, 36.

Heracleous, Loizos and John Murray (2001) 'The Urge to Merge in the Pharmaceutical Industry', *European Management Journal*, 19(4), 430–7.

Hollensen, Svend (2004) *Global Marketing: A Decision-Oriented Approach*, 3rd ed (Harlow: Pearson Education).

Hoppenstedt Firmeninformationen GmbH (2003) *ASTA Medica GmbH*; accessed at *www.lexisnexis.co.uk*.

Howell, Julian (2004) 'The Why and How of US Market Entry: A Qualitative Study of Non-US Pharmaceutical Companies', *International Journal of Medical Marketing*, 4(3), 235–50.

Jones, Trefor T. and Cockerill, T. A. J. (1984) *Structure and Performance of Industries* (Oxford: Philip Allan).

Kanavos, Panos and Elias Mossialos (1999) 'Outstanding Regulatory Aspects in the European Pharmaceutical Market', *Pharmacoeconomics*, 15(6) (June), 519–33.

Kay, John (1993) *Foundations of Corporate Success: How Business Strategies Add Value*, (Oxford University Press).

Kettler, Hannah (2001) (ed.) 'Introduction' in H. E. Kettler (ed.), *Consolidation and Competition in the Pharmaceutical Industry*, based on papers delivered at the OHE Conference, London, 16 October 2000 (London: Office of Health Economics).

Langley Amanda, Nada Korac-Kakabadse and Stephen Swailes (2005) 'Grand Strategies and Strategic Actions in the Pharmaceutical Industry 2001–2002', *Technology Analysis and Strategic Management*, 17(4), 519–34.

Lawrence, Peter (2002) *The Change Game: How Today's Global Trends are Shaping Tomorrow's Companies* (London: Kogan Page).

LEK (2004) '*Company Overview*', accessed at *www.lek.si/eng/company-overview/about-lek*.

Li, Jiatao (1995) 'Foreign Entry and Survival: Effects of Strategic Choices on Performance in International Markets', *Strategic Management Journal*, 16(June), 333–51.

Lindblom, Charles (1959) 'The Science of Muddling Through', *Public Administration Review*, 19(Spring), 79–88.

Lindblom, Charles E. (1979) 'Still Muddling, Not Yet Through', *Public Administration Review*, November/December, 517–25.

Lynch, Richard (1997) *Corporate Strategy* (London: Pitman).

Mason, E. S. (1959) *Economic Concentration and the Monopoly Problem* (Cambridge, Mass.: Harvard University Press).

Matraves, Catherine (1999) 'Market Structure, R&D and Advertising in the Pharmaceutical Industry', *The Journal of Industrial Economics*, XLVII, (June), 169–92.

McKelvey, M., L. Orsenigo and F. Pammolli (2004) 'Pharmaceuticals Analysed through the Lens of a Sectoral Innovation System', in F. Malerba (ed.), *Sectoral Systems of Innovation* (Cambridge University Press), 73–120.

Miles, Matthew D. and Michael Huberman (1994) *Qualitative Data Analysis: An Expanded Sourcebook*, 2nd edn, (London: Sage Publications).

Mintzberg, Henry, B. Ahlstrand and J. Lampel (1998) *Strategy Safari: A Guided Tour through the Wilds of Strategic Management*, 3rd edn, (New York: The Free Press).

Patton, Michael Q. (2002) *Qualitative Research and Evaluation Methods*, 3rd edn, (Newbury Park, Calif.: Sage).

Pearce II, J. A. (1982) 'Selecting Among Alternative Grand Strategies', *California Management Review*, XXIV(3), 23–31.

Pearce II, J. A. and R. B. Robinson Jr, (1994). *Strategic Management Formulation, Implementation And Control*, 5th edn (Homewood, Illinois: Irwin).

Pierre Fabre (2003) *Pierre Fabre 2002 Annual Report* (Castres Cedex: Pierre Fabre).

PJB Publishing (2001) *Scrip Pharmaceutical League Tables* (Richmond, Surney: PJB Publishing).

Porter, M. E. (1980) *Competitive Strategy Techniques for Analyzing Industries and Competitors* (New York: The Free Press).

Pursche, William R. (1996) 'Pharmaceuticals–the Consolidation Isn't Over', *The McKinsey Quarterly*, (2), 110–19.

Quinn, James B. (1980) *Strategies for Change: Logical Incrementalism* (Homewood, Ill.: Irwin).

Quinn, James B. (1991) "Strategic Change: 'Logical Incrementalism'", in J. B. Quinn and H. Mintzberg *The Strategy Process: Concepts, Contexts and Cases*, 2nd edn (London: Prentice-Hall International (UK) Ltd).

Rugman, Alan M. and Cecilia Brain (2004) 'The Regional Strategies of Multinational Pharmaceutical Firms', *Management International Review*, 44(7), pp. 7–25.

Scherer, F. M. (1980) *Industrial Market Structure And Economic Performance*, (2nd edn, (Chicago, Rand McNally College Publishing Company).

Schmidt, S. and Edwin Ruhli (2002) 'Prior Strategy Processes as a Key to Understanding Mega-Mergers: The Novartis Care', *European Management Journal*, June, 223–34.

Scrip (1992a) 'Asta's New Pharma Plant in Dresden', 27 March (Richmond Surney: PJB Publications), 9.

Scrip (1992b) 'Galen Expands with New Antibiotics Plant', 17 June (Richmond Surney: PJB Publications), 15.

Scrip (1994) 'Pierre Fabre Foundation', 11 October, p. 13 (Richmond Surney: PJB Publications).

Scrip (1998) 'Pierre Fabre Expanding Global Presence', 10 July, (Richmond Surney: PJB Publications), 11.

Taggart, James (1993) *The World Pharmaceutical Industry* (London: Routledge).

Tolentino, P. E. (2000) *Multinational Corporations: Emergence and Evolution* (London: Routledge).

Walsh, V. and I. Galimberti (1993) 'Firm Strategies, Globalisation and New Technological Paradigms', in M. Humbert (ed)., *The Impact of Globalisation On Europe's Firms and Industries* (London: Pinter).

Walton, J. (2001) 'Investors' Views on Merger and Acquisition, Alliance and Licensing Activity in the Pharmaceutical Industry', in H. E. Kettler (ed.), *Consolidation and Competition in the Pharmaceutical Industry* (London: Office of Health Economics).

Webb, David and Andrew Pettigrew (1999) 'The Temporal Development of Strategy: Patterns in the U.K. Insurance Industry', *Organization Science*, 10(5), 601–21.

9
Multinational Companies' Battle against Counterfeiting

Elfriede Penz

Introduction

Counterfeiting has been spreading across the globe at an alarming rate in almost all industry sectors, making up about 9 per cent of world trade, or USD450 billion a year, in 2002 (IACC, 2002). While the phenomenon occurs worldwide, in developing countries the combination of large profit margins and low observation of the law, together with light penalties and low compliance with international regulations, lead to even higher rates of counterfeiting (Shultz and Saporito, 1996).

Business sectors such as luxury goods, pharmaceuticals, video and audio recordings, fast-moving consumer goods, textiles, toys or software are mainly affected by counterfeiting (ICC, 2004; Markenverband, 1999). Exclusive products will always be counterfeited, but a growing number of counterfeiters are tending to focus on low-margin mass-produced goods rather than on quality high-margin luxury goods, increasingly generating profits from quantity rather than quality. One of the benefits of this trend for counterfeit sellers is that fake mass-produced goods are less likely to be detected at border controls (*The Economist*, 2003).

The purpose of this chapter is, first, to review companies' actions as recommended in the literature. Second, an empirical investigation of the companies' application of these actions aims at surveying the experiences and where possible their effectiveness in reducing counterfeiting. Finally, the theoretical and empirical evaluation provides the basis for managerial recommendations for multinationals to curb counterfeiting.

Conceptual background

Manufacturers of original products are well aware of counterfeiting and follow suggestions to limit damage to their company's brand reputation and profits (see, for example, Green and Smith, 2002; Kay, 1990; Nash, 1989; Wee *et al.* 1995).

Network theory can be used as a framework to examine the relationships MNCs have with various political actors (Hadjikhani and Ghauri, 2001; Hadjikhani and Sharma, 1999) in order to protect their brands. In this chapter, business political behaviour is viewed as 'actions taken to favourably position the firm in its nonmarket environment by managing those uncertainties and resource dependencies stemming from the influence and/or resistance of other nonmarket actors that (can) affect the firm's overall economic performance' (Salorio *et al.*, 2005, p. 30). This view transcends the confines of economic relations with environments by assuming that a firm positions itself in non-market environments.

Failures in the context of counterfeiting can derive from malfunctioning relationships – for example, with suppliers, distributors or competitors, lack of information or lack of access to governmental or institutional actors. Partnering has become a vital element in internationalization and is viewed as an increasingly important competency to reduce risks in international business (Cavusgil, 2006).

In the following, an overview is provided of management attempts to cope with counterfeiting. Existing literature mainly reflects business activities to curb counterfeiting. According to insights from the discussion on political actors, the following section reviews relations that companies have with non-market actors. Companies' activities were classified regarding the relationship between actors, resulting in (i) business-to-consumer relations (b-to-c); (ii) business-to-business relations (b-to-b); and (iii) business-to-government (b-to-g) relations.

Business-to-consumer relations

Harvey and Ronkainen (1985) stress the importance of increasing the goods' status appeal by registering or numbering products, or by having only a few company-owned retail outlets. Methods of marking brands and thus facilitating the identification include raised lettering on packaging, special ink and dyes, or genetic markers (Bush *et al.*, 1989; Shultz and Saporito, 1996). Labelling the genuine product with high-tech marks might be a good start to fight off counterfeiters, but in most cases pirates use modern technology to fake goods and are able rapidly to copy the applied high-tech labels. Companies constantly change and evolve brands by improving performance, or redesign the packaging and labels (Bush *et al.*, 1989; Chaudry and Walsh, 1996; Shultz and Saporito, 1996). Those customers who willingly buy counterfeits are made aware that only original products are superior (Shultz and Saporito, 1996). Yet, if an affected luxury brand company were to adopt a warning strategy, there is a risk that consumers might choose brands from its competitor (Chaudry and Walsh,

1996). Thus, communication with customers needs to be tailored according to the industry, and expected damages to the brand.

Business-to-business relations

These can even involve the 'bad guys' by legitimizing criminal business. This action is effective within countries where it is difficult to control counterfeiters because of geographical size, number of inhabitants, variety of languages and lack of co-operation with local authorities – for example, China and Thailand (see Green and Smith, 2002).

'Doing nothing' follows a basic cost–benefit analysis. Factors such as market size, cost associated with litigation and enforcement, insignificant levels of damage to authentic goods, the company's brand name and image, or short product life-cycles are reasons why companies could decide to keep their hands off the problem (Harvey and Ronkainen, 1985; Shultz and Saporito, 1996). Another rationale for ignoring piracy is that piracy is not all bad and helps in diffusing products (Prasad and Mahajan, 2003). Distributors are usually the first to come into contact with fake goods that are smuggled into the distribution line (Bush *et al.*, 1989; Harvey, 1987). In order to identify counterfeits, distributors need to be trained to raise their awareness and to judge the authenticity of the merchandise.

Investigation and surveillance activities suggest that companies hire investigators or establish in-house detectives. Their task is to monitor the market by buying samples in various shops, and in the case of a counterfeit, following the line of distribution back to the manufacturer (Bush *et al.*, 1989; Harvey, 1987; Harvey and Ronkainen, 1985; Jacobs *et al.*, 2001; Shultz and Saporito, 1996).

Business-to-government relations

These involve both economic and political parties. Companies protect their goods by using existing patent laws, trademark laws or brand protection laws by registering their brand names at the respective patent offices. However, protection of trademarks and copyrights is not simply a matter of less developed countries (Harvey, 1987; Harvey and Lucas, 1996; Shultz and Saporito, 1996). Often a single firm may not have the necessary resources to curb counterfeiting. Thus companies establish coalitions and associations, or a single company joins an existing association with similar intellectual property rights (IPR) interests. Various associations have been established since the mid-1980s, aiming at building a platform for exchange of information and lobbying for conjoint concerns. Yet, 'the basic problem is that all companies would like to have more information but few wish to contribute' (Vithlani, 1998, p. 38). Anti-counterfeiting associations exist at both a *global* level (for example, the Counterfeiting Intelligence Bureau (CIB), the Global Anti Counterfeiting Group (GACG)) and a *national level* (for example, Anti-Counterfeiting Group

(ACG) and the Brand Protection Association (BPA)). In order to meet business-specific concerns, several *company-driven associations* have been established (for example, the Business Software Alliance (BSA), and the Imaging Consumables Coalition of Europe (ICCE)).

Companies pursue political approaches that aim, directly or indirectly, at governments through international organizations, European Union (EU) regulation or free trade agreements (Bush *et al.*, 1989; Chaudry and Walsh, 1996; Harvey, 1987). From a global perspective, the World Trade Organization (WTO) plays a crucial role in the protection of intellectual property rights – for example, in China (Hung, 2003).

The above-mentioned actions have been labelled according to the lead in the affair – 'proactive', 'reactive' or 'protective' measures (Harvey and Ronkainen, 1985; Jacobs *et al.*, 2001). No empirical evidence of these actions exists, nor any coherent framework for an anti-counterfeiting strategy. By conducting a series of interviews with managers, plus company-related documents from various sources (newspapers, organizations and so on), this chapter aims at introducing such a framework from a network perspective. Based on the literature and existing concepts of counterfeiting, research questions were developed to guide the research process. The questions concerned views on counterfeiting, and actions against it, with an emphasis on relations with consumers, business and the government.

Methodology

A qualitative approach was applied using N*Vivo (Richards, 2000; Richards, 2002; Richards and Richards, 1994). The grounded theory approach is the basic methodology for the present study (Strauss and Corbin, 1994; Strauss and Corbin, 1998). Care was taken to control for research bias – that is, taking into account different response patterns and physical conditions in qualitative interview settings to establish equivalence of research. Two research assistants were trained to approach the respondents and conduct interviews. Their way of interacting, and their experiences and insights were carefully supervised and discussed. The aim was to assess maximum external validity by contrasting managers' responses from different company sectors.

Overall, managers from sixty multinational companies were contacted, by telephone, e-mail, post or at events. Industry-wise, they represented fast-moving consumer goods (FMCG), pharmaceuticals, software, luxury goods, textiles and sportswear, and consumer electronics. The business press was used to identify companies that were faced with counterfeiting. At the end, six out of the sixty companies were willing to collaborate in the study, and interviews with eight managers were conducted. The low response rate was caused by the sensitive topic of the survey. Three of the interviewed managers were responsible for marketing and PR (sports and luxury goods; FMCG), the manager for the software company was head of distribution in small to medium enterprises, and two FMCG managers were business unit managers. The companies were all multinationals, with headquarters in Germany, France, the Netherlands, and the USA. They each employ between 9,500 and 165,000 employees worldwide. The interviews were conducted by two interviewers, and were carried out in the offices of the managers, apart from two telephone interviews, which were conducted from university offices equipped with telephone conference and tape-recording facilities.

Data analysis

Data analysis consisted of organizing steps, coding, and searching procedures (Richards, 2002; Richards and Richards, 1994; Sinkovics *et al.*, 2005).

Organizing steps

Construct validity was established by three tactics proposed by Lee (1999) and Yin (2003) – that is, using multiple sources of evidence, establishing a chain of evidence, and providing feedback to respondents. Interviewees' comments, research assistants' observations during the interview setting, and contextual factors were integrated into the project database as well as information from company- and organization websites and from the business press (*Financial Times, The Economist* and so on). Researchers confirmed the relevance of the collected information to the overall research questions by rating the sources according to the concept being studied. The comments were discussed and a common rating was established for the operational measures used. This procedure assured a logical, sequential process by following the same process (contacting the manager, conducting the interview, writing up observations) for each interview, which allowed for reconstruction and anticipation by external audit. At the end, the multiple elements of information converged and indicated high construct validity. A protocol was developed from the procedure to realize repeatability and replicability of the procedures. The managers were contacted once again, a summary of the interview was provided and their comments requested. To guarantee replicability, data were digitized, tape-recorded and transcribed in their original language. Five of the six interviews were held in German; and one telephone interview was held in English. The original language was retained, and therefore the transcripts were in both German and English.

Coding processes

The obtained project database was the basis for coding, a process that is best understood as the ongoing interpretation of the data from as many angles as possible. A-priori and a-posteriori categorization of data was applied. *A-priori categorization* involved the development of categories prior to the actual data collection, based on theory and literature review. *A-posteriori categorisation* came after the data collection and was based on the given responses.

Each interview was analysed applying three coding strategies – namely, open, axial and selective coding. *Open coding* was used for the discovery of categories and identification of new concepts. *Axial coding* applied categories and concepts to the empirical data. Categories were related to their subcategories, and intersections of related categories were identified. The objective of axial coding was to add depth to the categories that had been created by open coding. The outcome of axial coding was a node system consisting of six main nodes and numerous sub-nodes, resulting in 105 nodes in total. *Selective coding* means integrating and refining categories in order to build a theory. Certain notions were highlighted and relevant statements used to explain the phenomenon of interest in order to reach a certain level of abstraction.

Search processes

The next step was to theorize about the content by conducting various search processes. The main search processes aimed at understanding managers' general knowledge and insights into counterfeiting – that is, what are counterfeits, and the issue of superiority of originals with respect to counterfeits. Further search processes were conducted to gain an understanding of the actions and to develop anti-counterfeiting strategies.

Results

General knowledge and insights into counterfeiting

Managers from the FMCG sector reported that the majority of products were only partly counterfeited – for example, the brand name, but not identical layout (colours), or vice-versa. Cases of counterfeits with similar brand design were also reported. It seems that there exists not just one kind of counterfeit in the FMCG sector but diverse imitations. As managers from the sports apparel and the luxury goods firm, they have found a basis on which to judge whether a product is a counterfeit or not. Only those parts of technical products that were not protected by patents were deliberately changed against rebuild parts reflecting the know-how necessary to fake electronic goods. In a business sector in which patents have become the basis for success, each violation or threat is monitored by companies. For the representative of a software company, there existed two forms of pirated copies: those that are easily recognizable and those that are not. The representative reported the case of an original product split into several subproducts, combined with fake product parts, making more products out of one. A manager reported what he called 'casual copying' – that is, receiving the software from friends, and 'excessive use' of licences – that is, using one licensed software package on more than one computer. These represented the most common threats to software companies.

Asian countries were perceived to be the most experienced and the most technically advanced producers of counterfeits. Eastern European countries that had recently joined the European Union (EU) were not considered to be sufficiently technically advanced. However, as in many Western European countries, counterfeits appeared in the former Soviet bloc countries in great numbers. For the latter, it was assumed that their borders were rather weak, and therefore counterfeit products could easily enter the EU.

All of the interviewed managers agreed that globalization, entering new markets and converging consumer needs and preferences are intensifying the phenomenon of counterfeits. Becoming more international was regarded as a reason for increased threats through counterfeits (sports apparel, FMCG).

Not only products with a high image potential have been counterfeited but also such items as flagstone glue or wallpaper paste. In Western Europe, DIY consumers spend more money on accessories such as glue and so on, but Eastern Europeans usually save on such products, making the availability of cheap and fake adhesives more attractive there.

Although counterfeits appear worldwide, certain patterns seemed to emerge when looking at different product categories with regard to country characteristics. Fake functional products emerged almost everywhere in the world, but differences exist in buying behaviour. In Western European countries, consumers sometimes unknowingly buy fakes when, for example, competitors use the design of an original product to sell their counterfeit brands. Southern European countries such as Italy, Spain or Turkey were regarded as hot selling spots for this activity. The EU seemed to be a transit area rather than a main target for counterfeits. However, counterfeits appear more frequently and more systematically in developing countries, where demand is much higher because of limited purchasing power and better possibilities of getting the counterfeits on to the shelves or sold on the open market.

The interviewed managers knew quite well where counterfeits were being produced – usually the Far East, in particular China. One case was reported in which products were shipped from China to the USA, where they were sold through large, well-established supermarket chains. Countries around the Black Sea (especially Ukraine with its port in Odessa) and Russia, but also North Africa, were mentioned as counterfeit-producing countries. Within Europe, only Poland and Turkey were mentioned by the interviewees.

Counterfeiters, figuratively called 'the pirates' by one interviewee, were assumed to belong to organized groups, even to the mafia. They also had connections to the authorities, such as the police or customs. In particular, Russia, Ukraine, Bulgaria and the Far East were countries with high levels of organized counterfeiting activities and connections to other forms of crime. In one case, counterfeits were identified through the inspection of a plant where children were being employed illegally, which links the activities of counterfeiters to child labour.

Most of the managers knew about smuggling routes and believed that counterfeits entered countries via the same routes as did drugs. They assumed that counterfeiters planned their actions rigorously and even calculated for loss through seizure. Some of the best-known geographical routes were from the Far East to Switzerland, and to Odessa in the Ukraine. The borders of the new Eastern European member states provided easy access to smuggle counterfeits into the EU.

Activities against counterfeiting

The actions that were identified through literature review were the basis for the analysis (a-priori categorization). Three actions were not considered by the interviewed managers, namely leaving the industry, doing nothing and integrating counterfeiters. The strongest resistance was found against doing nothing. Managers explicitly mentioned that 'one cannot simply do nothing' to deal with the problem (see Figure 9.1).

With respect to country-specific approaches, a software manager mentioned that similar actions were not effective everywhere: the Business Software Alliance (BSA), for example, is very successful in some countries, but not in countries

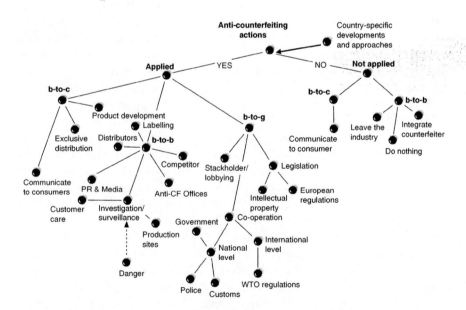

Figure 9.1   Categorization of company activities against counterfeiting

where the BSA is not well established. Therefore, the authority of an association has an impact on the success of a strategy including various partners. In the case of the consumer electronics company, the patent offices collaborate very well. Whenever a counterfeit product that mimics an original appears in the Far East, the respective offices on site take over the investigations.

In general, actions were adapted to respective national requirements. A FMCG company, for example, is tailoring its efforts according to the markets and rarely applies the same action across markets: 'We try to identify the best solution, taking into account the...type and size of the problem for the country concerned.' The consumer electronics manager pointed to the fact that, especially in the area of intellectual property, 'one has to look ahead and observe how a country is developing'.

Business-to-consumer

Managers conceived consumers to be largely innocent with respect to counterfeit purchase. In some cases, companies had direct contact with consumers who brought a (counterfeit) product to a store in order to have it repaired, or in the case of software, to get it authorized. Managers thought that consumers had simply not recognized the counterfeit as such and largely ignored these cases.

To understand the problem better, companies have set up call centres or web-based feedback systems to offer their customers the opportunity to report incidences of counterfeits or just to complain. In the technical goods sector, consumers complained about malfunctions of products because of a poor fit with accessories (such as CDs or DVDs). Only after having contacted the manufacturer, did these customers recognize that they had bought a counterfeit product. With respect to the FMCG sector, consumers complained about malfunctions in the products (detergents or adhesives). Counterfeits were sold in normal shops, making it almost impossible for consumers to know the origins of the goods. In the case of luxury brands, buyers of original products reported places where they had identified counterfeits on sale. They complained and expected the company to deal with the problem. On this topic, the manager from the luxury goods sector observed that consumers would refrain from buying original luxury products because of the large number of counterfeits on the market. With respect to software, the interviewed manager said that honest customers strongly believed in justice and therefore reported incidences. Some companies encouraged this by offering rewards.

Companies' representatives described both affluent consumers and average earners as potential counterfeit purchasers. Two representatives (of luxury and sports goods) assumed that consumers act according to normative pressures from their environment. Fashion trends and the way that information about brands was communicated influenced consumer desire to obtain the brand, either a genuine version or a copy. There is even more pressure when only few pieces are available and the particular product is perceived as a 'must-have'. Where there were delivery problems, customers who could not wait simply bought counterfeits.

Managers from the apparel industry (luxury and sports) believed that consumers buy counterfeits instead of the original because they want to be considered attractive and smart by peers. One manager stated that counterfeit handbags could be used every day, whereas an original one was reserved for special occasions.

Based on their insights from customer reports and the generally poor quality of counterfeits, managers feared that counterfeits would harm the company image. Once counterfeits appear on the market, this means fading trust. Managers were especially concerned about the effects of unknowingly purchased counterfeits: 'And so you will say that [product xy] or another brand is a lousy product! And so you will also tell your friends and you won't buy it again, and it will be very difficult to get you back to the product. So that's why it is so important' (FMCG manager).

There were different reactions to the question of whether to inform consumers about the problem or not. The manager from the software company thought it was necessary to make consumers aware of copied software. One of the FMCG managers agreed that the company had to inform customers about incidences in affected countries. The other two managers in the same business segment were reluctant to tell consumers about the problem as were managers from the luxury and sports segments. The latter two felt that it would harm the company's image rather than help them to curb the problem.

Control over distribution was one way that was applied by luxury goods and sports apparel companies. Offering products via exclusive distribution channels was expected to control the problem and guaranteed that all goods that were bought in normal shops or from authorized dealers would be originals. While counterfeits were more likely to be sold outside normal and authorized stores, this was not true in developing countries, where fake products appear on store shelves.

Business-to-business

In developing countries, companies employ agencies to monitor counterfeits. Success stories included closing down counterfeiting plants, confiscating counterfeit products at borders/in shops, recovering damages from trademark violations, and criminal prosecutions. Yet, these were only marginal efforts: 'And I hope to some extent we have it under control ... it will probably never go away completely.' Cutting down on counterfeiting was regarded only as a short-term success. By tracking sales and customer complaints, managers can tell whether counterfeits have reappeared on the market. It seems that, within the company, it is very important for managers to demonstrate successful measures against counterfeiting.

Within companies there exist databases detailing incidences of counterfeits, reports on legal actions, booklets, and material for training distributors, customs and police. The information remained confidential and for internal usage only, such as for keeping track of a number of identified counterfeits, sales numbers and customer feedback. A FMCG manager reported that the company has a specific knowledge management system for that kind of data, while in the luxury goods company internal information was rather scarce and employees were not very well informed. Some companies (luxury and sports goods, software) organized internal meetings or conferences, at which the issue was discussed at the international level.

One important action managers took to protect their products and brands was to have them registered. No matter what business the company was in, this included all possible visible and unseen parts of the product or brand, and their design. Some companies' trademarks are also registered by other parties. For example, ten years ago companies did not protect their trademarks in China because the economy was too weak and they did not expect the country ever to enter the market. Some Chinese companies took advantage of this, registered the brands as their own and can now produce goods legally, as has happened in the consumer electronics industry.

Software was also protected by copyrights, and infringement of these can incur severe penalties. Patent and design rights were actively protected by the consumer electronics manufacturer.

Three of the interviewed managers said that they had special departments dealing with IPR. The managers from the luxury goods and consumer electronics companies also referred to their internal anti-counterfeiting offices. The advantage of these was that the companies could act more independently and have direct control. Companies had engaged private detectives to find out where the counterfeits were produced, from which countries they came, and, if produced outside the European Union how they entered the EU.

All the interviewed managers mentioned business partners or private investigators who identified production sites, trade routes or groups of counterfeiters. Employees of the sports goods company were themselves responsible for identifying counterfeits. However, since counterfeiters mainly acted within organized criminal groups, the danger of such investigations was acknowledged. The FMCG manager mentioned a retailer in the Ukraine who was badly injured in a bomb attack because he had informed the police.

Companies with a luxury product portfolio usually remain silent about instances of counterfeits that they discover. In the case of the FMCG sector, the media was informed about arresting dealers and so on in order to let the public as well as potential counterfeiters know that such companies were active.

All the interviewed managers stated that their products were protected by technical or other security features, which were changed regularly. It was clear to them that consumers must not be able to detect changes to the packaging or label, because they would then expect the product to have changed too. As the software representative and one FMCG manager agreed, 'basically, it's a race; a race between the one who holds the rights and those who counterfeit them'. Relations with distributors, wholesalers, and retailers was very important, as another FMCG manger stated: 'We basically try to act on the supply chain-level of the product, before it reaches the customer. Of course, that is not always possible.'

Distributors were trained to recognize any deviation from the original product, and were rewarded if they reported them. This worked well for products that were sold at exclusive or reliable stores in developed countries. In developing countries, fake goods were sold in normal shops, and distributors were in a difficult position when companies detected counterfeits and held *them* responsible: 'We have programmes where we have been focusing only on the distribution of our products to make sure that the ones who get caught with the fake products do not get ... products any more' (FMCG manager). Within the FMCG sector, and in the sports goods company, collaboration with competitors seemed to be move important than in the other sectors (for example, the luxury goods sector). In the affected countries, the notion of competition lost its meaning: 'We do everything on an industry basis. So, we are all in the same boat and we are not in competition on counterfeits. We fight together against counterfeits in these countries' (FMCG manager).

Business-to-government

Companies depend largely on governmental structures such as the legal system, or compliance with regulations that were imposed on a global level, and relations with police and customs. Although not all interviewed managers knew exactly to which association their company belonged, they agreed that collaboration with the authorities at both national and international level proved to be successful. The national associations were regarded as quite practical groups.

Custom officials were perceived as being important partners because of their powerful positions. In particular, in developed countries where counterfeits were sold but not produced, customs control helped to identify counterfeit goods. The police were perceived as being equally important, their role being to prosecute counterfeiters. Finally, collaboration with governments in developing countries was mentioned, and in the case of FMCG a clear connection with organized crime was established: 'And it seems more and more, we see that ... governments have become more active because counterfeits are an, let's say, easy way for organized crime to make a lot of money.'

In addition to registered trademarks, patents and copyrights, company managers perceived relevant legislation to be the most important tool to reduce counterfeiting. For all of them, the application of laws and regulations was clear and they reported several cases of raids, penalties and legal actions. The number of these actions was perceived as the most tangible measure of success.

New European regulations and their authority over the new member states were perceived very positively by a manager who was aware of the pan-European effect on national legislation. The new EU members would benefit by introducing strong intellectual property protection laws. To ensure that governments attached the necessary priority to the problem of counterfeiting, to raise awareness and to seek support, company representatives (FMCG, consumer electronics) had meetings with members of governments.

Finally, the companies can be positioned according to their major problem markets (developed versus developing markets) and their major anti-counterfeiting activities (see Figure 9.2). While fake FMCG appeared mainly in developing countries, consumer electronics struggle with counterfeits in developed countries. Both industries focus on establishing governmental ties in the form of legislation, lobbying and stakeholder education. Luxury and sports goods manufacturers deal with fakes in developed countries by setting up business-to-business relations – for example, by engaging detectives. Finally, counterfeit software and similar illegal forms appear in various markets, but companies react with customer-orientated anti-counterfeiting strategy.

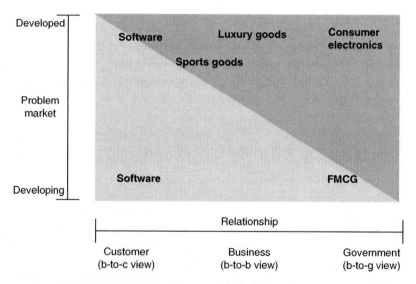

Figure 9.2 Positioning of studied companies with regard to problem markets and category of anti-counterfeiting activities

Discussion

Counterfeits are a big threat to companies, resulting in reduced turnover, damage to brand image and high commitment in juridical activities. This chapter focused on relations and studied actions that have been applied by multinational companies to deal with counterfeiting. Activities were identified that are based on (i) business-to-consumer relations (b-to-c); (ii) business-to-business relations (b-to-b); and (iii) business-to-government relations (b-to-g).

However great is the problem of counterfeiting, there are doubts about the extent of the problem from an economic perspective. In a recent article on software piracy, *The Economist* pointed to the fact that the economic impact of software piracy is difficult to calculate (*The Economist*, 2005); others, including Microsoft's Bill Gates, hold the opinion that piracy should be tolerated (Prasad and Mahajan, 2003). Interviewed managers also agreed that a company's biggest costs are those finding out how large the problem is, and continuously monitoring the extent.

The study revealed that a great deal of information existed within companies that was helpful in tracing counterfeiting. Customer complaints could be used systematically as an alert system to identify cases of counterfeit products, as well as the commitment of loyal customers who deliberately report incidences of fakes. Providing rewards for such information might help in the long run not only to retain customers but also to provide secure originals. Monitoring the choice behaviour of both groups of consumers – those who buy counterfeits and those who opt for the original – could also inform the company about its marketing strategies, whether it is an unacceptable pricing policy or a promoted high image that cannot be reached realistically by the majority of consumers.

Our findings suggest that companies perceive counterfeiting to affect all, or many, firms, suggesting that collaborative efforts may be worthwhile. Partnering is a critical competence when operating in international markets, and will gain in importance. From the internationalization perspective, anti-counterfeiting strategies should begin very early – that is, at the print where the decision is made as to where and when to enter a new market. Struggles with counterfeiting in the Far East have shown that the mode of entry is a critical managerial decision and, as was proved in China, where distribution is not nationally co-ordinated, counterfeiters can easily outrun the original manufacturer.

Knowledge about patent and copyright regulations and membership of associations might reduce the danger of patent or copyright infringement. Indonesia, for example, holds legal patents and copyrights for Western brands and therefore makes it – legally – impossible to avoid 'fake' products being sold in those markets. Additionally, the position of a potential market regarding WTO regulations proves useful is learning about potential threats.

A major threat in entering markets in the Far East is the cultural idea of product development. As outlined by Hung (2003), China's product development strategy is one of replicating and imitating. As long as these concepts prevail and are in contrast to Western companies' ideas, counterfeiting will remain an almost unsolvable issue. Although Green and Smith (2002) were able to show that there is a chance of establishing relations with counterfeiters, the information obtained from the managers in this study point to the reverse being the case.

There are several limitations to the present study. First, considering the extent of the problem, the number of managers willing to participate was not very high. In a subsequent study, therefore, a better response rate will prove beneficial to include details of as many experiences as possible. Second, managers were basically talking to outsiders, and probably not providing information about sensitive internal data that might have helped to understand fully a company's strategy. A challenge for future research is therefore to obtain a broad picture of the phenomenon. Third, the outcome of companies' strategies was not related entirely to successful measures; thus future research should incorporate details of successful measures of anti-counterfeiting activities.

References

Broughton, Martin F. (2003) 'Counterfeiting: A New Business Risk', *World Economic Forum*, http://www.weforum.org (accessed 17 March 2004).

Bush, Ronald F., Peter H. Bloch and Scott Dawson (1989) 'Remedies for Product Counterfeiting', *Business Horizons*, 32(1), 59–65.

Cavusgil, Tamer (2006) 'International Business in the Age of Anxiety: Company Risks, Keynote speech, Academy of International Business (AIB-UK), Manchester, UK.

Chaudry, Peggy E. and Michael G. Walsh (1996) 'An Assessment of the Impact of Counterfeiting in International Markets: The Piracy Paradox Persists', *Columbia Journal of World Business*, 31(3), 34–48.

Creswell, John W. (1997) *Qualitative Inquiry and Research Design: Choosing among Five Traditions* (London: Sage).

Green, Robert T. and Tasman Smith (2002) 'Countering Brand Counterfeiters', *Journal of International Marketing*, 10(4), 89–106.

Hadjikhani, Amjad and Pervez N. Ghauri (2001) 'The Behaviour of International Firms in Socio-political Environments in the European Union', *Journal of Business Research*, 52(3), 263–75.

Hadjikhani, Amjad and D. Deo Sharma (1999) 'A View on Political and Business Actions', *Advances in International Marketing*, 9, 243–57.

Harvey, Michael G. (1987) 'Industrial Product Counterfeiting: Problems and Proposed Solutions', *Journal of Business and Industrial Marketing*, 2(4), 5–13.

Harvey, Michael G. and Laurie A. Lucas (1996) 'Intellectual Property Rights Protection: What MNC Managers Should Know about GATT?', *Multinational Business Review*, 4(1), 77–93.

Harvey, Michael G. and Ilkka Ronkainen (1985) 'International Counterfeiters: Marketing Success without the Cost and the Risk', *Columbia Journal of World Business*, 20(3), 37–45.

Hung, C. L. (2003) 'The Business of Product Counterfeiting in China and the Post-WTO Membership Environment', *Asia Pacific Business Review*, 10(1), 58–77.

IACC (2002) *International AntiCounterfeiting Coalition, 2002* (Washington, DC: International Anti-Counterfeiting Coalition).

IACC (2005) *The Negative Consequences of International Intellectual Property Theft: Economic Harm, Threats to the Public Health and Safety, and Links to Organized Crime and Terrorist Organizations* (Washington, DC: International Anti-Counterfeiting Coalition).

ICC (2004) *The International Anti-counterfeiting Directory 2004*, (Barking: ICC).

Jacobs, Laurence, A. Coskun Samli and Tom Jedlik (2001) 'The Nightmare of International Product Piracy', *Industrial Marketing Management*, 30(6), 499–509.

Kay, Helen (1990) 'Fake's Progress', *Management Today*, July, 54–8.

Kenrick, Douglas T., Norman P. Li and Jonathan Butner (2003) 'Dynamical Evolutionary Psychology: Individual Decision Rules and Emergent Social Norms', *Psychological Review*, 110(1), 3–28.

Lee, Thomas W. (1999) *Using Qualitative Methods in Organizational Research* (Thousand Oaks, Calif.: Sage).

Markenverband (1999) 'Counterfeiting', *http://www.apm.be* (accessed 15 February 2004).

Nash, Tom (1989) 'Only Imitation? The Rising Cost of Counterfeiting', *Director*, May, 64–9.

Prasad, Ashutosh and Vijay Mahajan (2003) 'How Many Pirates Should a Software Firm Tolerate? An Analysis of Piracy Protection on the Diffusion of Software', *International Journal of Research in Marketing*, 20, 337–52.

Richards, Lyn (2000) *Using NVivo in Qualitative Research*, 2nd edn (Bundoora, Victoria, Australia: QSR International Pty).

Richards, Lyn (2002) *Introducing NVivo: A Workshop Handbook*, Vol. 2004.

Richards, Thomas J. and Lyn Richards (1994) 'Using Computers in Qualitative Research', in Norman K. Denzin and Yvonna S. Lincoln (eds), *Handbook of Qualitative Research* (Thousand Oaks, Calif.: Sage).

Salorio, Eugene M., Jean Boddewyn and Nicolas Dahan (2005) 'Integrating Business Political Behavior with Economic and Organizational Strategies', *International Studies of Management & Organization*, 35(2), 28–55.

Shultz, Clifford J., II and Bill Saporito (1996) 'Protecting Intellectual Property: Strategies and Recommendations to Deter Counterfeiting and Brand Piracy in Global Markets', *Columbia Journal of World Business*, 31(1), 18–28.

Sinkovics, Rudolf R., Elfriede Penz and Pervez Ghauri (2005) 'Analysing Textual Data in International Marketing Research', *Qualitative Market Research: An International Journal*, 8(1), 9–38.

Strauss, Anselm and Juliet Corbin (1994) 'Grounded Theory Methodology – An Overview', in Norman K. Denzin and Yvonna S. Lincoln (eds), *Handbook of Qualitative Research* (Thousand Oaks, Calif.: Sage).

Strauss, Anselm L. and Juliet M. Corbin (1998) *Basics of Qualitative Research: Grounded Theory Procedures and Techniques* (Thousand Oaks, Calif.: Sage).

The Economist (2003) 'Imitating Property Is Theft – Counterfeiting', *The Economist*, 367.

The Economist (2005) 'Business: BSA or just BS? Software Piracy', *The Economist*, 375.

Vithlani, Hema (1998) *The Economic Impact of Counterfeiting* (Paris: Organization for Economic Co-operation and Development – OECD).

Wee, Chow-Hou, Soo-Jiuan Tan and Kim-Hong Cheok (1995) 'Non-price Determinants of Intention to Purchase Counterfeit Goods', *International Marketing Review*, 12(6), 19–46.

Yin, Robert K. (2003) *Case Study Research: Design and Methods*, 3rd edn (Thousand Oaks, Calif.: Sage).

10
Expanding the International Business Research Agenda on International Outsourcing

Jussi Hätönen and Mika Ruokonen

Introduction

As diminishing national barriers, improved communications links and the evolution of a new, focused supplier base has made it possible to move internal activities and processes across corporate and national borders, the proponents of international business literature have become increasingly interested in the phenomenon of international outsourcing. Ramamurti (2004) has stated that outsourcing across national borders is the new and expanding topic of future international business (IB) research, but as yet the research community has paid only limited attention to this important phenomenon. Accordingly, in this chapter we shall illustrate how several aspects of the phenomenon have been overlooked by IB scholars. We shall not only emphasize the importance and need for further research on the topic, but also to identify some of the important future issues for IB scholars regarding future topics on international outsourcing.

Outsourcing is not yesterday's phenomenon, although interest in this topic has rocketed in recent years. Several different streams of concurrent business literature have contributed greatly on the general practice of outsourcing. For example, previous studies from the strategic management perspective have examined outsourcing from angles such as a tool for restructuring organizations into more flexible forms (for example Lei and Hitt, 1995) – that is, transformational outsourcing (for example, Linder *et al.*, 2002); the motives towards outsourcing (for example, Kakabadse and Kakabadse 2002); the value of outsourcing (for example, Quinn and Hilmer, 1994); the pitfalls of outsourcing (for example, Barthélemy, 2003b); effects of outsourcing on the firm's performance (for example, Gilley and Rasheed, 2000) and market value (for example, Hayes *et al.*, 2000); how to manage outsourcing relationships (for example, Useem and Harder, 2000) to name but a few. Recently, as requested by Ramamurti (2004), much more focus has been given to outsourcing in the international context (for example, Beulen *et al.*, 2005; Doh, 2005; Farrell, 2005; Levy, 2005) covering issues such as what to outsource and where (Graf and Mudambi, 2005; Palvia, 2004) and recently the topic has been attracting increasing interest among IB scholars.

Despite this growing interest of researchers in the topic, however, the most important question has not yet been fully answered: that is, why do some fail and others succeed in their outsourcing arrangements? Contradictory findings over the applicability of outsourcing of similar organizational processes and functions to similar destinations suggest that the answer to this question lies beyond 'what' and 'where'. Companies seeking to outsource internationally face adversities, often related to the risks of operating in international markets. On the other hand, in a turbulent business environment, outsourcing can diminish company risks, whether they are commercial- or country-related. However, several risks exist in outsourcing itself, such as knowledge spillovers through outsourcing core elements, or for the hollowing of organizations losing control of operations. Nevertheless, it has been stated that, if done correctly and for the right reasons, outsourcing can provide unseen benefits, ones that might assist greatly in operating in today's turbulent markets.

The purpose of this chapter is to raise new issues and aspects concerning the timely topic of international outsourcing, ones that may have been previously overlooked, subconscious, or entirely new. To be able to pinpoint areas for further research from the IB perspective, current research on international outsourcing and its connection to other streams is first analysed – that is, seeking to answer the question 'How has the IB research stream contributed to overall research on outsourcing?' In relation to the identified research gaps we shall illustrate three case studies that together provide a vivid tool to illustrate the practical connection of the suggested possibilities for further research directions in international outsourcing.

As can be seen from Figure 10.1, the study illustrated in this chapter was constructed from three phases whose purpose was to create implications for future research on international outsourcing (Phase 4). First, we conducted a literature review to gain an understanding of current research interest in the topic. Second, we analysed fifteen companies to discover their outsourcing practices and experiences. The purpose of these case studies was to refine and reinforce the current theories, to suggest new research areas on international outsourcing and possibly create hypotheses for further research. Third, based on the literature review and the company analyses, we suggest gaps and thus future directions for international outsourcing research. As already noted, to emphasize the practical proximity of these issues, three examples from the company analyses connected to these issues are provided in the chapter.

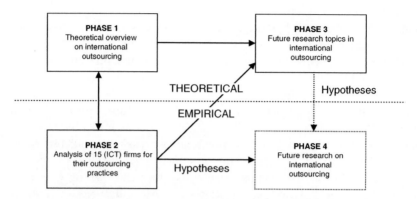

Figure 10.1   Outline of the chapter and its purpose for future studies

Identifying the gaps in international outsourcing research

Although the topic of international outsourcing is often examined in an interdisciplinary manner, Maskell *et al.* (2005) have been able to identify three types of literary stream with an interest in international outsourcing, each with its own characteristics: strategic management literature, supply chain literature and international business (IB) literature. We could introduce an additional fourth stream of literature that has existed since the beginning of the practice, and that is on information systems/technology (IS/IT). However while IS, or IT, literature often draws on strategic management literature, IB literature and sometimes supply chain literature, the issues behind this stream derive from the three main streams of outsourcing literature only in a specific industry context. Thus, in this chapter we shall only introduce the three main streams and their contribution to the understanding of offshore outsourcing[1]. A summary of current research as relevant research equations in and stream is represented in Table 10.1.

First, *strategic management literature* is largely focused on the resource base, core competencies and boundaries of the firm, and is thus focused more on explaining the phenomenon of outsourcing in general rather than offshore outsourcing, although recent developments in that field are leaning towards explaining economic and managerial incentives for offshore outsourcing.[2] However, the main focus of the strategic management literature is to seek the rationales behind actions regarding outsourcing. From the perspective of strategic management, outsourcing was previously seen as a tool to trim the cost base (Coase, 1937; Williamson, 1975) or to acquire resources that were in short supply or that were not available internally (Johanson and Mattsson, 1988; Pfeffer and Salancik, 1978). At the time, outsourcing was seen as a tool to trim

Table 10.1 Current research streams on international outsourcing and their special characteristics

Research stream	Main questions addressed	Some recent contributors	Discipline-based theories
International business literature	• Why outsource abroad?	Farrell, 2004a; 2005; Quélin and Duhamel, 2003	Geographical location theory, transaction cost theory, resource-based view
	• Where to outsource?	Palvia, 2004; Graf and Mudambi, 2005; Vestring *et al.*, 2005	Geographical location theory
	• What is the impact of offshore outsourcing on different stakeholders?	Deavers, 1997; Baily and Farrell, 2004; Farrell, 2004b; Doh, 2005; Levy, 2005	Institutional theory
	• Is offshore outsourcing an incremental learning process?	Hagel and Brown 2005; Graf and Mudambi 2005; Maskell *et al.*, 2005; Beulen *et al.*, 2005	Internationalization process theories
Strategic management and supply chain literature	• What should be outsourced?	Baden-Fuller *et al.*, 2000; Gilley and Rasheed, 2000	Internalization theory (transaction cost theory, resource-based view)
	• Who is the right supplier for outsourced activities?	Hoetker, 2005; Feeny *et al.*, 2005	Network approach
	• How to manage outsourcing relationships?	Barthélemy, 2003a; Lonsdale, 1999; Useem and Harder, 2000; Håkanson and Ford, 2002	Relationship theory, network approach
	• Vertical versus virtual integration?	Fine *et al.*, 2002; Lawton and Michaels, 2001; Bagchi and Skjoett-Larsen, 2002	Theory(-ies) of the firm, organization theory, internalization theory
	• How to manage the new dispersed supply chains?	Lonsdale, 1999; Parker and Russell, 2004	Relationship theory, network approach

Sources: Main contributing sources are Buckley and Lessard, 2005; Maskell *et al.*, 2005.

organizational efficiency through handing out non-critical peripheral functions to specialized providers. However, and as a result of increased competition alongside falling interaction and communications costs and the evolution of specialized suppliers, companies are moving towards outsourcing their more critical functions and processes. Simultaneously, the focus has shifted from strict cost discipline to creating superior customer value. Furthermore, the distinction between core and non-core competencies is fading, and some researchers have even touched upon the issue of outsourcing the core competencies (Baden-Fuller *et al.*, 2000; Gilley and Rasheed, 2000).

This increased outsourcing activity has, according to some management scholars, generated a platform and need for dynamic set of new core competencies. The first of them can be referred to as strategic restructuring competence. The essence of this is that a company's real value in creating competency, perhaps its only sustainable one, might even accumulate from its ability to restructure its value chain continuously (see, for example, Fine *et al.*, 2002). Another 'new' core competence discipline arises from the company's ability to manage the geographically dispersed network of suppliers resulted from the outsourcing of economic activities (Kakabadse and Kakabadse, 2002), also referred to as network competence (Gemünden and Ritter, 1997). Accordingly, the strategic management literature on outsourcing is increasingly focusing around three main questions: (i) what could or should not be outsourced; (ii) for those activities that are outsourced, what the right supplier would be; and (iii) for those activities that are outsourced, what form of relationship is most appropriate? International aspects of outsourcing set challenges for strategic management research related to the latter two questions and provides possibilities for the first one.

The second stream of international outsourcing literature is *supply chain literature*, in which outsourcing is looked at from the value chain and distribution perspective, such as procurement. Vertical integration versus virtual integration has been one of the main comparative analyses in this stream (Bagchi and Skjoett-Larsen, 2002), which often draws on theories such as internalization theory, organization theory and different theories of the firm (Buckley and Lessard, 2005). According to Bagchi and Skjoett-Larsen (2002), it is costly and even sometimes impossible to develop competitive capabilities in all areas, thus firms should identify areas where they can develop or acquire capabilities of their own, and where they can rely on supply chain partners to provide the required competitive capabilities. For example, along with global or multi-vendor sourcing arrangements, supply chains are becoming global and more complex and, as companies are demanding just-in-time deliveries or in-line sequences, issues such as logistics and procurement have become one of the mainstays of the outsourcing literature (Ellram and Billington, 2001). Although configuring downstream operations has been the main focus of both outsourcing and offshore outsourcing in the supply chain literature, innovative business models have brought a need for new capabilities in upstream supply chain management. Thus the outsourcing literature is increasingly concentrating also on that area (Lawton and Michaels, 2001).

Finally, *international business literature* is emphasizing international localization and factor aspects in explaining to same extent which outsourcing is conducted abroad. Research in international business and management has developed two broad traditions: variance theories and process theories (Langley, 1999). Variance theories aim to explain determinants of variation in corporate performance or behaviour, and typically are tested with cross-sectional data, whereas process theories seek to explain how and why businesses evolve over time (Meyer and Gelbuda, forthcoming).

From the IB perspective, offshore outsourcing literature has focused almost solely on developing variance theories. For example, the effect of offshore outsourcing has been studied in terms such as companies' performance (for example, Gilley and Rasheed, 2000) and market value (for example, Hayes *et al.*, 2000), but most importantly the IB literature has focused on certain stakeholder impacts such as on workers and labour (for example, Deavers, 1997), governments (for example, Baily and Farrell, 2004; Farrell, 2004b), NGOs (for example, Venkatraman, 2004) and societies in general (for example, Doh, 2005; Levy, 2005). However, from the variance perspective, much more of the IB scholars' attention should be paid to answering the question, 'Why do some companies fail while others succeed in outsourcing internationally', because successful outsourcing strategies are found to carry positive effects not only for the companies involved, but also for the stakeholders around them (for example, Farrell and Agrawal, 2003).

The variance studies on international outsourcing dominate the IB literature and only limited research has so for emphasized the process aspect of IB research. However, it has been noted that companies are increasingly outsourcing more and more critical aspects of their businesses abroad (Beulen *et al.*, 2005). Some researchers have recently started to scrutinize this evolution of offshore outsourcing. In the mid-1990s, for example, Quinn and Hilmer (1994) viewed outsourcing as a development process proceeding from short-term to long-term supplier contracts. Furthermore, Hagel and Brown (2005), for example, state that once a company has developed outsourcing skills, it is more likely to consider moving its outsourcing relationships to companies offshore. This type of incremental learning in outsourcing and offshore outsourcing is a growing topic of research today. For example, Graf and Mudambi (2005), as well as Maskell *et al.* (2005), state that offshore outsourcing is, or normatively should be, a sequential learning process in which cost advantage motives precede differentiation advantages, and near-shore locations precede far-shore outsourcing. Interestingly, this process is very similar to the early models of the outward internationalization process (Johanson and

Vahlne, 1977; Luostarinen, 1979). However, it is as yet unclear whether outsourcing might even have a positive effect on the outward internationalization of companies. Nevertheless, in accordance with this development, we should expect to come across the concepts of 'born global outsourcers' or 'international new outsourcing ventures' in the future.

Combining practice and theory – new aspects for IB research in international outsourcing

The question of 'what determines the international success and failures of firms' has always been the leading question guiding IB research, and will continue to be so in the twenty-first century (Peng, 2004). International business researchers have, however, to large extent neglected to examine outsourcing as a factor behind the success. IB scholars have been preoccupied with justifying the practice of outsourcing to stakeholders on a macro level, while the micro or industry level analyses of outsourcing and its effect on the overall success of companies has been left with inadequate attention. For example, in the field of IB, it has been shown to some extent that companies concentrating on their core competencies and outsourcing other activities are able to internationalize rapidly and with lower costs (Barthélemy, 2003b). Whereas strategic management literature has for a long time emphasized the implementation of a correct outsourcing strategy as one of the key factors behind the successes of modern companies, IB scholars still seem to look elsewhere for factors behind international success. In this chapter we shall introduce a few new perspectives of international outsourcing that have assisted companies in unconventional and unrealized ways in their international business. As a result, we shall identify some issues of international outsourcing that could provide fruitful ground for future IB studies.

Explaining international firms through outsourcing

Along with reducing national barriers, improved communication links and the evolution of a new, focused supplier base, companies are starting to outsource some of their international operations, as the ever-changing markets shift location and internalization advantages. This causes entirely new management issues within multinational corporations, and yet these issues have been overlooked by IB scholars. Furthermore, there exist only limited studies on what causes companies to take on-shore once-outsourced operations. In this age of anxiety and constant change, adding these aspects of international outsourcing to the concurrent IB research agenda could provide a better understanding of the future developments of the outsourcing phenomenon.

New aspects of international outsourcing through changes in the OLI paradigm

The common definition of outsourcing, which states that before something can be outsourced it has to be produced internally, constitutes the fact that companies cannot internationalize outwards directly through offshore outsourcing, although some authors (for example, Gilley and Rasheed, 2000) suggest that, in addition to the normal way, outsourcing can also occur through abstention, separate from basic procurement, because it only occurs when the internalization of the good or service outsourced was within the acquiring firm's managerial and/or financial capabilities. In short, in this approach, outsourcing is viewed as choosing to buy over make, if make was also possible with internal resources. However, this additional view is often seen as being too broad, as it confuses the practice of outsourcing with procurement and sourcing.

To a large extent, offshore outsourcing can be explained through the OLI paradigm in the way that ownership and location advantages exist, but not the internalization advantages (see Graf and Mudambi, 2005). In other words, it is useful to produce the product or service in the foreign location, but not with internal resources. The only way in which outsourcing is an international operation mode as referred to by the internationalization process models, is changes in the OLI paradigm, and in particular in the advantages in internalization. (See Table 10.2.)

Dunning (1988, p. 63) predicted that:

> MNEs would wish to reduce their presence in a particular country or sector under two circumstances. First, where a change in the distribution of factor endowments (or the efficiency with which these are used) (1) weakens their competitive advantages, relative to those of firms in host countries, or (2) causes them to switch production from the host to home (or indeed, other host countries). Second, where the net transactional benefits (costs) or using the external markets for the exploitation of these competitive advantages increase (fall) relative to those offered administered hierarchies.

Table 10.2 Extension to the routes of serving markets

Route of serving markets	Ownership	Advantages internalization	(Foreign) location
Foreign direct investment	Yes	Yes	Yes
Trade in goods and services	Yes	Yes	No
Contractual resource transfers	Yes	No	No
Offshore outsourcing	**Yes**	**No**	**Yes**
Changes			
Outsourcing foreign direct investment	Yes	Yes → No	Yes
Offshore outsourcing of foreign direct investment	Yes	Yes → No	Yes → No
Relocating outsourced activities	Yes	No	Yes → No

Source: Dunning, 1988.

What Dunning stated to be one of the emerging issues is such divestment and/ or relocation of international activities. Several internationalization process models (for example, Johanson and Vahlne, 1977; Luostarinen, 1979), Dunning (1988) state that companies might seek to divest themselves of their international operations in time, yet remain present in the markets through non-equity arrangements such as partnerships, which on the other hand equates with operating abroad through an outsourcing arrangements.

The changes in the internalization and location advantages provide a fruitful and important ground for further studies in the field of IB. In fact, it could be argued that the whole topic of international outsourcing arises from the changes in the advantages of the OLI paradigm. However, many of the possible scenarios are still unrealized and thus provide possibilities for further research. For example, the late proponents of the 'staged models' of internationalization have acknowledged that the internationalization process of a company can in fact move it down the ladder as well as up, but future studies on international outsourcing might illustrate why. The following case illustrates an example of how changes in location and internalization advantages are shaping a company unit from being a foreign production unit towards being a foreign sales unit.

Outsourcing international operations: case company A

Company A is a computer-aided design (CAD) software producer providing the manufacturing industry as well as the building and construction industries with a highly sophisticated but standardized product. The company has its headquarters in Finland and sales subsidiaries in the USA, Singapore, the UK and France. The US subsidiary was established at the start of the 1990s. In the US market the company operates solely with the product for building and construction. This product is sophisticated, and is also therefore somewhat complex, so it takes time for the customer to learn how to use it. This is why the product is often sold with a two-week on-site training course, and a two-year update and customer services package is often added to the final product. When the company entered the US market the entire solution was provided with internal resources. However, after the US business expanded, the company reached a situation where its internal resources, both time and money, were inadequate to serve such a package on such a large continent. First, they found that internalization of the on-site training was not profitable as it tied up the scarce time of software developers, and because there existed a competent supplier in that field to provide that service to them. Furthermore as this relationship developed, company A found that the provision of after-sales service was not profitable and it was better to outsource that as well, partially to the same supplier. The outsourcing decisions which resulted from the changes in OLI advantages over time are represented in Table 10.3.

Table 10.3 Changes in company A's OLI advantages

Services of the CAD product	Ownership	Advantages internalization	(Foreign) location
Establishment of subsidiary A (early 1990s)	Yes	Yes	Yes
Partial outsourcing of implementation services to a US-based provider (mid 1990s)	Yes	Yes → No	Yes
Partial outsourcing of after-sales services to a US-based provider (late 1990s)	Yes	Yes → No	Yes
Relocating after-sales operations (India, 2006?)	Yes	No	Yes → No

Now, however, the company is facing a situation where the location-specific advantages have changed in comparison to those in India, for example. Falling interaction costs and the development of know-how in developing countries has made it more attractive for companies to outsource low-level service jobs to these countries. Thus the company's US subsidiary is now considering relocating some of the outsourced services, such as after-sales call centres, to low-cost locations. Along with further improvements in information and communication links, it might be possible in the future that, for example, India's location advantages precede those of the USA even with regard to implementation services.

The effect of international outsourcing on the outward internationalization process

The internationalization of companies is one of the mainstays of IB research. The proponents of the 'staged models' of internationalization (for example, Johanson and Vahlne, 1977; Luostarinen, 1979) suggest that, as companies acquire more foreign-market knowledge, the more commitment decisions are made. Furthermore, it was argued that outsourcing is an incremental learning process. However, in this chapter we shall argue that, international outsourcing is also an incremental learning processes, and might in fact have an effect on a company's outward internationalization process.

International outsourcing as an incremental learning process and horizontal–outward connections

According to Hagel and Brown (2005) companies do not immediately outsource the most critical activities abroad. In fact, recent studies suggest that off-shore outsourcing is a sequential learning process in which cost advantage motives precede differentiation advantages, and near-shore locations precede far-shore outsourcing (Graf and Mudambi, 2005; Maskell *et al.*, 2005). Sequentality and incrementality suggest that companies gradually engage in outsourcing first in home markets and only after that do they seek providers from off-shore markets to provide, first, perhaps less critical functions, but later even the most critical ones (see Figure 10.2). Thus location-specific advantages such as productivity, quality, availability of resources, infrastructure and so on are not sufficient to explain the off-shore outsourcing of non-core activities of the firm (Dunning, 1988), but bounded rationality also affects international outsourcing decisions.

Although some studies have been conducted based on the fact that companies move from outsourcing first peripheral activities and then move towards more strategic ones, outsourcing first in near-shore locations, and moving incrementally to far-shore locations, more studies could be done on linking these two separate but similar processes. This is illustrated in Figure 10.2. It is to be expected that companies acquire market knowledge from the outsourcing country. However, it is unclear whether this market knowledge can be transferred. Prior research suggests that the success of outward internationalization is dependent on the effective use of inward internationalization, and vice versa. Furthermore, success behind the internationalization of a company might lie in its previous outsourcing of activities. It is interesting that the inward type of internationalization has received only limited attention, and even more so when some studies have found that inward internationalization strongly affects the outward internationalization process (for example, Welch and Luostarinen, 1993). However, these studies have concentrated almost solely on the procurement and sourcing aspect of inward internationalization, leaving many aspects of the issue of outsourcing untapped.

The basic idea of off-shore outsourcing affecting companies' outward internationalization process is illustrated in Figure 10.3. The underlying

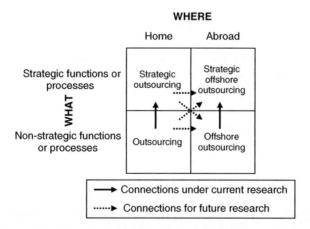

Figure 10.2   The incremental processes towards strategic offshore outsourcing

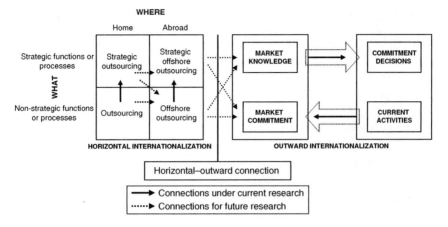

Source: Adapted from Johanson and Vahlne, 1977.

Figure 10.3   Horizontal–outward connections through outsourcing

assumption of this model, that a gradual increase in off-shore outsourcing will increase the market knowledge in a particular market, has been seen as a prerequisite and one of the decisive factors in an outward internationalization process (Johanson and Vahlne, 1977). Furthermore, the international outsourcing of activities can *per se* increase the market commitment, and this in turn could explain the jumping-over-phases phenomenon in companies' internationalization processes.

The following case illustrates a situation in which a company outsources solely to acquire market knowledge. Although on a very narrow margin as marketing and sales is partially outsourced in this case, it illustrates a situation where an outsourcing agreement is used to boost the outward internationalization process of the company. Based on the following case, it is possible to expect that under some conditions, market knowledge of the target market can be acquired and transferred through outsourcing arrangement.

Outsourcing for market and industry intelligence: company B

Company B is also a Finnish-based design software producer. The company can be considered to be truly global, as today almost 95 per cent of its sales accumulate from abroad, from seventy different countries around the world. Furthermore, the company is represented in fifteen countries, as their product is often sold as a solution that also includes a service aspect, mainly in the form of training and after-sales services such as updating and a support centre.

The company realized that for a sales organization, which company B considers itself to be, there can be no sales without understanding the markets. In particular, they realized this fact as they decided to expand on to the Japanese market. After setting up a subsidiary, the company encountered several problems, from simple language difficulties to the established methods of conducting business. The Japanese markets turned out to be more complicated than had been expected, and thus the company divested a part of its representation in those markets and simultaneously made a partnership agreement with a local reseller to cover the gap left by the divested part. Thus in a way they outsourced a piece of their local representation or sales to a more competent unit. The main reason and motive behind this decision was to acquire knowledge about the complicated Japanese market. They retained office of their own on the markets so that they would be able to work in close co-operation with the reseller. Even though the company hardly breaks even in Japanese markets today, they still pursue this strategy as they see that the market knowledge that is acquired through the partner could be useful in future efforts to increase commitment on to those markets.

The network effect of outsourcing

In addition of gaining market knowledge, international outsourcing can also assist in other ways the international success of companies. Through international outsourcing a company can also gain valuable access to foreign networks that could eventually be extremely beneficial. In this chapter we shall justify why, in future international outsourcing research, the network effect should not be overlooked.

Internationalization through offshore outsourcing – network embeddedness and social networks

The process-based internationalization view has received a fair amount of criticism. One of the most common of these attacks the issue that in the traditional approach to internationalization, companies have been considered to be clearly-defined decision units with internally controlled resources (Andersen *et al.*, 1997). However, it is stated increasingly that the internationalization of a company is highly related to the context in which it operates (Madsen and Servais, 1997) and, more importantly, to the external resources available in each operational context (see, for example, Bonaccorsi, 1992). The process-based internationalization doctrine overlooks the value and the effect of the network in which the company is embedded (Holmlund and Kock, 1998).

The effect of outsourcing on a company's internationalization process can be further examined from the network perspective. As the company network arguably increases when it decides to outsource activities or processes so does it increase the possibility of its effect on the company. And it has been found that individual contact networks can play a crucial role in the internationalization process of a company (Axelsson and Agndal, 2000). This kind of individual contact network can be created through offshore outsourcing. On the other hand, individual contact networks can also lead to further offshore outsourcing commitments. Either way, the created network through (offshore) outsourcing often has an effect on the success of a company. In fact, a created downstream network can even result in direct influences on the sales of a company, as illustrated below in the case of company C. Thus, when seeking the factors behind the international success of companies, offshore outsourcing should also be examined from the network perspective.

Off-shore outsourcing as a springboard for outward internationalization: company C

Company C is a European-based software company whose sole business (in a diversified company) is to produce and license computer-aided design (CAD) software, mainly for construction and engineering offices. The firm's strengths lie in its knowledge and experience in providing customers with a highly-developed 3D modelling programme for building and construction, and that is where its core competencies lie.

As the construction industry is evolving beyond CAD to fully computerized projects, this has made new demands on software producers. The builders or designers are no longer satisfied with depicting the construction or the buildings, but are now demanding software programs that, for example, calculate the quantities of the elements that are needed, and work out in which part of the process they will be needed. These kinds of attributes can be developed through a separate calculation tool.

However, for a long time the company's capability to develop such a calculation tool were limited. As they wanted to focus their entire resource pool on the core product, 3D modelling, they decided to outsource the calculation part of the software solution. Accordingly, the outsourcing decision was a typical resource-based rationale. Because using multiple providers to supply them with this part of the solution would have meant building several interfaces to their product, they decided that the most cost-efficient way to proceed would be through a preferred partnership. After a search they settled on a California-based vendor that has specialized in this narrow and specific area.

As the relationship got off the ground the supplier soon realized that the performance of company C was having an effect on its own performance, and to some extent vice versa, so the Californian based supplier became a retailer of the combined solution in the US market (see Figure 10.4). Although the company already had sales in the US market before the retail agreement, this agreement provided company C with a further gateway to the large but troubled market in the USA.

What this case illustrates in short is a situation where outsourcing a part of the production process led to an increase in international sales – that is, outward internationalization. As a result of the decision to outsource the calculation part of the program, company C in fact gained unconscious and

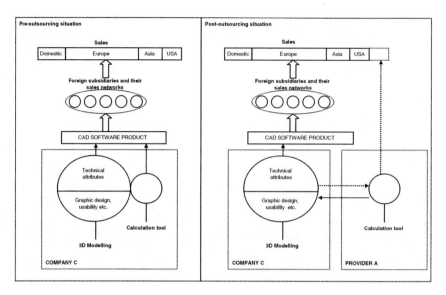

Figure 10.4   The horizontal–outward connection in company C

unplanned benefits. This case illustrates that it is important to take the target country's possible markets into account when making outsourcing decisions. If the company had decided to outsource to India, China or another low-cost location it might have found a cheaper price and perhaps even better quality, but it would have lost possible future sales. Thus, in future international outsourcing research, the network effect cannot be excluded from the analysis.

Conclusions and implications for future research

Since the birth of the practice, the research on outsourcing has gathered momentum, irrespective of a given literary stream. However, several aspects of this important topic have been overlooked or unrealized by the research community. The purpose of this chapter was to identify some of these issues from the IB standpoint. Accordingly, we first identified the scope and contributions of the IB research community in the wider outsourcing research field. Based on the IB literary overview provided in this chapter, we concluded that researchers should continue to seek an answer to the question of why some succeed and others fail in international outsourcing, as this question has not been answered fully in previous studies. In addition, we identified several issues that have been overlooked in previous research, which might provide a fruitful starting point for future IB research in the area of international outsourcing. According to our analysis, the following research questions and hypotheses could be introduced for further research:

- Under which conditions companies want to divest (a part of) their foreign presence?

  - $H_{1,1}$ Changes in the foreign internalization factors are behind the outsourcing decisions of foreign production.
  - $H_{1,2}$ International outsourcing of international operations (see Figure 10.1) is caused by simultaneous changes in location and internalization advantages.

    - $H_{1,2,1}$ Most often changes in location advantages precede changes in internalization advantages.

- Is outsourcing an incremental learning process, and how does international outsourcing affect the outward internationalization process of a company?

  - $H_{2,2}$ Previous domestic outsourcings have a positive effect on the outcomes of international outsourcing.
  - $H_{2,3}$ Previous outsourcing in a given country has a positive effect on the success of outward expansion to that country.

- How does international network embeddedness affect the international success of companies'?

  - $H_{3,1}$ International network embeddedness acquired through offshore outsourcing has a positive effect on the outward internationalization of companies.

These questions and hypotheses are examples of possible future research questions in international outsourcing. Testing them (Phase 4 in Figure 10.1) as such might create a new set of issues and problems. However, we shall not go any deeper into these hypotheses here, as the purpose of this chapter was not to test them, but only to identify them.

To summarize, generally more research is required on the aspects of international outsourcing other than outsourcing offshore – aspects such as outsourcing international operations or on-shoring operations. Scholars of IB should focus on scrutinizing this phenomenon from new perspectives, because, as demonstrated, international outsourcing can be the underlying factor behind the success of modern international companies. According to Peng (2004), this has always been one of the leading topics guiding international business research.

Notes

1. For a thorough literature review of outsourcing from the IS field, see Dibbern *et al.*, 2004, 35(4), 6–102.
2. For debate over the profitability of outsourcing, see *Journal of Management Studies* 2005, 42(3) (Farrell, 2005; Levy, 2005; Doh, 2005).

References

Andersen, Poul H., Per Blenker and Poul Christensen (1997) 'Generic Routes to Subcontractors' Internationalization', in Ingmar Björkman and Mats Forsgren (eds), *The Nature of the International Firm* (Copenhagen: Copenhagen Business School Press), 231–55.

Axelsson, Björn and Henrik Agndal (2000) 'Internationalization of the Firm. A Note on the Crucial Role of the Individual's Contact Network', Paper presented at the 16th IMP Conference, Bath, UK 7–9 September.

Baden-Fuller, Charles, David Targett and Brian Hunt (2000) 'Outsourcing to Outmanoeuvre: Outsourcing Redefines Competitive Strategy and Structure', *European Management Journal*, 18(3), 285–95.

Bagchi, Parabir K. and Tage Skjoett-Larsen (2002) 'Organizational Integration in Supply Chains: A Contingency Approach', *Global Journal of Flexible Systems Management*, 3(1), 1–10.

Baily, Martin N. and Diana Farrell (2004) 'Exploding the Myths about Offshoring', *http://www.mckinsey.com/mgi* (accessed 15 September 2005).

Barthélemy, Jérôme (2003a) 'The Hard and Soft Sides of IT Outsourcing Management', *European Management Journal*, 21(5), 539–48.

Barthélemy, Jérôme (2003b) 'The Seven Deadly Sins of Outsourcing', *Academy of Management Executive*, 17(2), 87–100.

Beulen, Erik, Paul Van Fenema and Wendy Currie (2005) 'From Application Outsourcing to Infrastructure Management: Extending the Offshore Outsourcing Service Portfolio', *European Management Journal*, 23(2), 133–44.

Bonaccorsi, Andrea (1992) 'On the Relationship between Firm Size and Export Intensity', *Journal of International Business Studies*, 23(4), 605–35.

Buckley, Peter J. and Donald Lessard (2005) 'Regaining the Edge for International Business Research', *Journal of International Business Research*, 36, 595–99.

Coase, Ronald H. (1937) 'The Nature of the Firm', *Economica*, 4(16), 386–405.

Deavers, Kenneth L. (1997) 'Outsourcing: A Corporate Competitiveness Strategy, not a Search for Lower Wages', *Journal of Labor Research*, 18(4), 503–19.

Dibbern, Jens, Tim Goles, Rudy Hirschheim and Bandula Jayatilaka (2004) 'Information Systems Outsourcing: A Survey and Analysis of the Literature', *The DATA BASE for Advances in Information Systems*, 35(4), 6–102.

Doh, Jonathan P. (2005) 'Offshore Outsourcing: Implications for International Business and Strategic Management Theory and Practice', *Journal of Management Studies*, 42(3), 695–704.

Dunning, John H. (1988) *Explaining International Production* (London: Unwin Hyman).

Ellram, Lisa and Corey Billington (2001) 'Purchasing Leverage Considerations in the Outsourcing Decision', *European Journal of Purchasing & Supply Management*, 7(1): 15–27.

Farrell, Diana (2004a) 'Beyond Offshoring: Assess Your Company's Global Potential', *Harvard Business Review*, 82(12), 82–90.

Farrell, Diana (2004b) 'Can Germany Win from Offshoring?', *http://www.mckinsey.com/ mgi* (accessed 15 September 2005).

Farrell, Diana (2005) 'Offshoring: Value Creation through Economic Change', *Journal of Management Studies*, 42(3), 675–82.

Farrell, Diana and Vivek Agrawal (2003) 'Offshoring and Beyond', *The McKinsey Quarterly*, 2003(4), 24–35.

Feeny, David, Mary Lacity and Leslie Willcocks (2005) 'Taking the Measure of Outsourcing Providers', *MIT Sloan Management Review*, 46(3), 41–8.

Fine, Charles, H. Vardan, Robert Pethick and Jamal El-Hout (2002) 'Rapid-response Capability in Value-chain Design', *MIT Sloan Management Review*, 43(2), 69–75.

Gemünden, Hans G. and Thomas Ritter (1997) 'Managing Technological Networks: The Concept of Network Competence', in Hans G. Gemünden, Thomas Ritter and Achim Walter (eds), *Relationships and Networks in International Markets* (Oxford: Elsevier).

Gilley, Matthew K. and Abdul Rasheed (2000) 'Making More by Doing Less: An Analysis of Outsourcing and Its Effects on Firm Performance', *Journal of Management*, 26(4), 763–90.

Graf, Michael and Susan Mudambi (2005) 'The Outsourcing of IT-enabled Business Processes: A Conceptual Model of the Location Decision', *Journal of International Management*, 11(2), 253–68.

Hagel, John and John S. Brown (2005) *The Only Sustainable Edge* (Boston, Mass.: Harvard Business School Press).

Håkansson, Håkan and David Ford, (2002) 'How Should Companies Interact in Business Networks?' *Journal of Business Research*, 55(2), 133–9.

Hayes, David C., James Hunton and Jacqueline Reck (2000) 'Information Systems Outsourcing Announcements: Investigating the Impact on the Market Value of Contract-granting Firms', *Journal of Information Systems*, 14(2), 109–25.

Hoetker, Glenn (2005) 'How Much You Know versus How Well I Know You: Selecting a Supplier for a Technically Innovative Component', *Strategic Management Journal*, 26(1), 75–96.

Holmlund, Maria and Sören Kock (1998) 'Relationships and the Internationalization of Finnish Small and Medium-sized Companies', *International Small Business Journal*, 16(4), 46–63.

Johanson, Jan and Lars-Gunnar Mattson (1988) 'Internationalization in Industrial Systems – a Network Approach', in Neil Hood and Jan-Erik Vahlne (eds), *Strategies in Global Competition* (London: Routledge), 287–314.

Johanson, Jan and Jan-Erik Vahlne (1977) 'The Internationalization Process of the Firm – a Model of Knowledge Development and Increasing Foreign Market Commitments', *Journal of International Business Studies*, 8, 23–32.

Kakabadse, Andrew and Nada Kakabadse (2002) 'Trends in Outsourcing: Contrasting USA and Europe', *European Management Journal*, 20(2), 189–98.

Langley, Ann (1999) 'Strategies for Theorizing from Process Data', *Academy of Management Review*, 24(4), 691–713.

Lawton, Thomas C. and Kevin Michaels (2001) 'Advancing to the Virtual Value Chain: Learning from the Dell Model', *Irish Journal of Management*, 22(1), 91–112.

Lei, David and Michael Hitt (1995) 'Strategic Restructuring and Outsourcing: The Effects of Mergers and Acquisitions and LBOs on Building Firm Skills and Capabilities', *Journal of Management*, 21(5), 835–59.

Levy, David L. (2005) 'Offshoring in the New Global Political Economy', *Journal of Management Studies*, 42(3), 685–93.

Linder, Jane C., Martin Cole and Alvin Jacobson (2002) 'Business Transformation through Outsourcing', *Strategy & Leadership*, 30(4), 23–8.

Lonsdale, Chris (1999) 'Effectively Managing Vertical Supply Relationships: A Risk Management Model for Outsourcing', *Supply Chain Management: An International Journal*, 4(4), 176–83.

Luostarinen, Reijo (1979) *Internationalization of the Firm* (Helsinki: Helsinki School of Economics).

Madsen, Tage and Per Servais (1997) 'The Internationalization of Born Globals: An Evolutionary Process?', *International Business Review*, 6(6), 561–83.

Maskell, Peter, Torben Pedersen, Bent Petersen and Jens Dick-Nielsen (2005) 'Learning Paths to Offshore Outsourcing – from Cost Reduction to Knowledge Seeking', DRUID Working Paper No. 05–17.

Meyer, Klaus E. and Modestas Gelbuda (forthcoming) 'Process Perspectives in International Business Research in CEE', *Management International Review*, forthcoming.

Palvia, Shailendra C. J. (2004) 'Global Outsourcing of IT and IT Enabled Services: A Framework for Choosing an (Outsourcee) Country', *Journal of Information Technology Cases and Applications*, 6(3), 1–20.

Parker, David W. and Katie Russell (2004) 'Outsourcing and inter/intra supply chain dynamics: strategic management issues', *Journal of Supply Chain Management*, 40(4), 56–68.

Peng, Mike W. (2004) 'Identifying the Big Question in International Business Research', *Journal of International Business Studies*, 35(2), 99–108.

Pfeffer, Jeffrey and Gerald Salancik (1978) *External Control of Organizations: A Resource Dependence Perspective* (New York: Harper & Row).

Quélin, Bertrand and François Duhamel (2003) 'Bringing Together Strategic Outsourcing and Corporate Strategy: Outsourcing Motives and Risks', *European Management Journal*, 21(5), 647–61.

Quinn, James B. and Frederick Hilmer (1994) 'Strategic Outsourcing', *Sloan Management Review*, 35(4), 43–55.

Ramamurti, Ravi (2004) 'Developing Countries and MNEs: Extending and Enriching the Research Agenda', *Journal of International Business Studies*, 35(4), 277–83.

Useem, Michael and Joseph Harder (2000) 'Leading Laterally in Company Outsourcing', *Sloan Management Review*, 41(2), 25–36.

Venkatraman, Venkat N. (2004) 'Offshoring Without Guilt', *MIT Sloan Management Review*, 45(3), 14–16.

Vestring, Till, Ted Rouse and Uwe Reinert (2005) 'Hedge Your Offshoring Bets', *Sloan Management Review*, 46(3), 27–9.

Welch, Lawrence S. and Reijo Luostarinen (1993) 'Inward-Outward Connections in Internationalization', *Journal of International Marketing*, 1(1), 44–57.

Williamson, Oliver E. (1975) *Markets and Hierarchies. Analysis and Antitrust Implications* (New York: The Free Press).

Part IV

SME Internationalization, Entrepreneurship and the Internet

11
Growth of a Greek International New Venture across Geographic Markets and Industries*

Pavlos Dimitratos, Irini Voudouris and Helen Salavou

Introduction

Although international entrepreneurship has been a field of emerging interest since the early 1990s, it recently shifted its emphasis to the study of opportunity identification. Previous attempts to demarcate the boundaries of this field have centred on the speed of internationalization as the key criterion distinguishing an international entrepreneurial firm (see, for example, McDougall, 1989; Oviatt and McDougall, 1994), yet recent writings increasingly emphasize the notion of exploitation of opportunity by the international entrepreneurial firm (Dimitratos and Plakoyiannaki, 2003; Oviatt and McDougall, 2005). Dimitratos and Jones (2005) assert that the theme of 'international opportunity perception' may form one of the most important and well-researched fields in international entrepreneurship study in the future. It is also strongly related to organizational growth because firms that perceive and act on opportunities abroad can grow successfully in the international marketplace. In spite of this, existing empirical evidence on this issue seems to be missing.

In particular, we appear to lack evidence on how entrepreneurial firms evaluate and act on opportunities that appear in the market. It could be that opportunities arise at home or abroad, and so examination of the pursuit of profitable prospects by the entrepreneur should take place everywhere the internationalized firm seeks opportunities, regardless of the locus of the geographical market and industry. Apart from this, it would be interesting to investigate how internationalized firms pursue opportunities to mobilise scarce resources in dynamic

* An earlier version of this chapter was written as a case study with the financial support of the European Case Study Writing programme of the Gate2Growth Academic Network in Entrepreneurship, Innovation and Finance.

environmental contexts that offer significant opportunities for growth. It is towards these objectives that we aim to provide some empirical evidence in this chapter.

We chose to carry out a case study on an entrepreneurial small firm in the information and communication technology (ICT) sector. Also, because researchers (for example Katsikeas *et al.*, 1997; Leonidou *et al.*, 2002) have often stressed that more studies should be conducted in countries other than the major industrialized ones, this current research offers insights from a small country with a dynamic economy located on the periphery of the European Union (EU), namely Greece.

This case describes the creation and analyses the growth of Alpha S.A., a Greek ICT service provider. The evidence illuminates the process by which the firm seeks to exploit profitable prospects in the marketplace, wherever they arise. Because the study covers a seven-year period, it highlights appropriately key factors affecting the pursuit of opportunities and growth in the marketplace of a dynamic environment. This evidence suggests additionally how ventures in growth environments can be managed successfully, thus offering useful guidelines to management practitioners. The chapter is structured as follows. The second section provides the research background to the theme of exploitation of opportunities and related growth in international markets. The third section outlines methodological details of the case study. The fourth section presents and discusses the findings of the research, and the concluding section highlights the key points of the study and discusses implications for theory and practice.

Literature review

An emerging theme in the entrepreneurship literature deals with the degree to which firms search and discover opportunities, evaluate opportunities among alternatives, and finally exploit those selected. Shook *et al.* (2003) assert that there are three stages in the opportunity perception process, namely intention, search and the discovery of opportunities. Compared with other internationalized firms, international entrepreneurial firms are likely to be more successful regarding how quickly, efficiently and holistically they sense and act upon opportunities abroad (see Crick and Spence, 2005). As far as international entrepreneurial intention is concerned, Shapero's (1982) model may be applied in international entrepreneurship studies. According to this model, entrepreneurial intentions are determined by perceptions of feasibility and desirability, and a propensity to act on an opportunity. In relation to international entrepreneurial search, awareness of opportunities abroad can be instrumental to initiation of entrepreneurial ventures. With regard to international entrepreneurial discovery, alertness to opportunities is of key importance because it can lead to the detection of attractive prospects in the international marketplace

(Minniti and Bygrave, 2001). Organizations with alert decision-makers are more likely to become successful entrepreneurs (Minniti, 2004). Alertness leading to the discovery of opportunities is an iterative process that depends on learning from organizational successes and failures (Gaglio, 1997). For example, organizations that value learning in their internationalization process can be more alert and in a better position to discover opportunities abroad than those who do not assign importance to learning.

There is a dearth of studies associated with the pursuit of market opportunities in the international entrepreneurship field. McDougall *et al.* (1994) argue that founders of firms that go abroad from their inception, namely international new ventures, are alert to new international market opportunities because of their previous knowledge and the learning acquired from the earlier activities of their founders. Thus the stock of knowledge that founders of international new ventures possess is influential to the fast speed of internationalization in these firms. In a similar vein, Zahra *et al.* (2000) argue that positive associations exist for international new ventures between international diversity and technological learning, and between technological learning and performance.

Apart from this, Autio *et al.*'s (2000) study provides interesting insights which suggest that young internationalized firms can learn faster than older ones, and react promptly to opportunities. Thus these authors posit that older firms may have developed learning impediments that block their ability to learn quickly in the international marketplace. This can be related to the findings that international new ventures employ fewer established routines than do older internationalized firms for augmenting and exploiting the organizational stock of knowledge (Kuemmerle, 2002).

The growth of the firm, which is related to effective identification and exploitation of opportunities, has been approached in the strategic management literature in a rather narrow context, whereby the firm has to choose between different market segments and products, mainly in the domestic market (Ansoff, 1965). The international business literature has also approached the growth of the firm as a predominantly foreign business activity, seemingly disregarding other 'domestic' or 'product market' strategic choices organizations may implement in order to discover opportunities and grow (see, for example, Johanson and Vahlne, 1977). It is surprising that the two literature streams have apparently not been examined in conjunction in order to provide a more holistic view of the overall strategic choice portfolio of the firm, particularly that of the small entrepreneurial company.

However, a few more recent contributions in the small-firm field acknowledge that small-enterprise internationalization is a holistic organizational phenomenon reflecting an internal growth and development process, in which both inward and outward internationalization interact with each other (Jones, 1999, 2001). In a similar vein, new venture strategies in developing countries may involve different combinations of product markets, technological capabilities and geographical target markets (Park and Bae, 2004). Hence, identification and exploitation of opportunities could take place in different product and geographical markets as the firm acquires experiential knowledge.

Methodology

A case study methodology was employed based on a successful small Greek firm, covering data over a seven-year period (1999–2005). Such longitudinal research can provide a dynamic view of the incident under investigation, providing in-depth insights into the theme of small-firm growth and exploitation of opportunities over a long time-frame (Yin, 1989; Patton, 2002).

The data collection methods used involved comprehensive interviews with a total of fifty-two managers, employees, partners and suppliers of the firm, the examination of company documents and archival data, and observation. Fourteen rounds of interviews were conducted at regular time intervals. In essence, the triangulation method was implemented, which is a procedure contributing to the acquisition of a comprehensive view and the generation of a valid interpretation of organizational phenomena. Collected data was content analysed. All interviews were transcribed. Initial respondents were asked to identify other managers and stakeholders of the firm involved in the process of growth of the company. Each interview lasted between one and one-and-a-half hours.

The interviews were based on a semi-structured questionnaire addressing key issues on the theme of firm growth, and opportunities identified and pursued in the company's markets. The themes of the questionnaire revolved around key milestones of the firm, as well as major facilitators and obstacles identified during the growth of the company. Respondents were asked to elaborate on how they reacted to potential opportunities identified, as well as how they dealt with major difficulties faced by the company in both international and domestic markets. As managers/employees in all departments of the firm and external partners/suppliers were interviewed, our insights were drawn from a broad range of value-added activities of the company, facilitating a comprehensive examination into the issue under investigation.

Findings and discussion

The early years of the firm

The company of which Alpha subsequently formed a part was founded in 1987 by various shareholders and the entrepreneur G.K., under the name Computer-Land R&D. From its early years the company targeted both domestic and international markets, offering consultancy, software development, software and testing. During its first nine years, the company grew slowly, although the

ICT market was developing rapidly. The two shareholders, namely ComputerLand and G. K, possessed equal shares of the company, yet they were not necessarily favouring the same decision when strategic issues were discussed. ComputerLand was more interested in the development of accounting software and tools, whereas G.K. was aware of the opportunities that could arise from the globalization of markets, thus leading to a high demand for software services.

Hence, it seemed logical for G.K. to take a management buy-out proposition to ComputerLand, and in 1996, G.K. became the owner of the company. Before starting to implement the first big change in the company, G.K. decided to reward and motivate his good colleague and friend G.S. by offering him, on preferential terms, 15 per cent of the company's shares. A deal was agreed and the two partners formed Alpha S.A., immediately implementing an aggressive business plan. The fact that the two partners were in essence the two main owners and strategists in this Greek firm is not unusual in the typical small firm, whose management style tends to be rather 'owner-controlled' and centralized. Family ownership, which is a feature of the vast majority of small Greek firms, may lead to a non-participative style of management in which the owners or top executives take all the key strategic decisions (Makridakis *et al.*, 1997; Voudouris *et al.*, 2000).

Reorganization and growth

With twenty people employed in the firm's early days, the top management team organised Alpha into two departments: the Department of I.T. and Telecommunication Services and the Department of Localization, Globalization and Quality Control. The newly-structured company targeted a market with significant potential. The global ICT market showed dramatic growth during the second half of the 1990s, with an average annual growth rate of 6.3 per cent for the years 1995–2001 (OECD, 2002). The software market was the most expanding subsegment of the ICT market, presenting an average annual growth rate of 12.3 per cent for the years 1992–2001.

The IT Department focused mainly on ICT outsourcing and solutions, a sector estimated to grow significantly in the future. Alpha's ICT outsourcing business line included a broad range of services, aiming at facilitating public and private (mainly) international organizations to optimize and control costs by allowing them to focus on their key competencies. Alpha offered end-to-end services covering the whole range of consulting, planning, deployment and management. In order to support this business unit, the firm participated and co-ordinated a considerable number of European Community (EC)-funded projects which were all successful, both in terms of scientific innovation and, most important, of the business applicability of the results. The participation in such demanding and innovative projects increased Alpha's technical expertise and capabilities for testing, localization and globalization. This expertise was also transferred to its other department. Furthermore, through these projects, the firm succeeded in developing a specialization in e-government services, which constituted the basis of the company's strategy in a later period. The firm developed its resources to such an extent that it was able to take advantage of later opportunities.

The Department of Quality Control, Localization and Globalization specialized in quality control and localization/globalization testing, translation and cultural adaptation of software and websites. Among the main offerings of the department were software testing and engineering services for international software producers. Starting from an early version of an original US software product, Alpha undertook all production and management tasks necessary to deliver ultimately a well-engineered and tested software product that met the high standards of its worldwide customers.

The overall market value of software testing and engineering services was difficult to measure, given that it was a niche market. However, an indicative example was that Microsoft spent between USD10–12 million per year in this market. Apart from Microsoft, new key customers such as IBM, which signed its first contract with Alpha in 1997, were attracted to the company, as they needed its specialized services in order to offer tested and localized versions of their products in the Greek market. Seeking new opportunities, an international new venture in 1999 offered localization services in other languages as well as Greek, including those of South-Eastern Europe and the Middle East. Existing customers supported this expansion because they were satisfied by the efficient services provided by the company, and were eager to further capitalize on its expertise. Alpha's expertise in localization was further strengthened through its co-operation with other European partners. The firm sought opportunities both abroad and at home, highlighting the point that the growth of a small firm can be a holistic organizational phenomenon (Jones, 1999), irrespective of its market locus.

G.K.'s persistence in discovering opportunities that lay in the localization industry was by then being shared by many others in the company, and the management team focused on localization, the importance of which, as a business activity (along with translation) was still unknown to many professionals at that time. G.K. was effective in transferring his vision for growth to many managers and employees within the firm, a characteristic frequently encountered in international new ventures (Madsen and Servais, 1997; Oviatt and McDougall, 1995). The localization/globalization market was showing signs of considerable potential: it was estimated that the total market would reach €1,380 million by the year 2001. Alpha's focus on the ICT service market sectors proved to be very successful. Synergies between the two departments helped the company to create unique capabilities in testing, localization and globalization. These technical capabilities, along with Alpha's emphasis on customer service and satisfaction were reflected in the company's annual growth rates. From 1997 to 2001, its turnover increased at an average annual rate of approximately 45 per cent. At the same time, the entrepreneur G.K. was always looking for new opportunities. The Athens Stock Exchange was booming at that time, and new prospects were arising for partnerships, buy-out schemes and so on.

Investment in e-goverment services – the creation of Alpha International Ltd

The increase in Alpha's turnover clearly indicated hyper-growth for the firm. This hyper-growth resulted in Alpha being selected as one of the twenty fastest-growing companies in Greece in 1999, and among the 500 fastest-growing companies in Europe in 2000. However, the two owners of the firm were aware of a potential problem that lay ahead: almost 70 per cent of the company's sales were coming from two big customers – namely, Microsoft and IBM. None the less, the fact that the investigated international new venture gravitated towards leading foreign customers is indicative of the internationalization behaviour of such 'knowledge-intensive' firms (Bell *et al.*, 2004).

Thus in 2001, once again, G.K. decided on a new internal transformation. Taking advantage of the technical and consulting expertise in the e-government area that the company had acquired through its involvement in relevant EU R&D projects in previous years, the Department of I.T. and Telecommunication Services focused strategically on e-government, an area were many opportunities were emerging. This was because, under the European Commission's e-Europe initiative, the year 2005 was set as the deadline for all member states to offer the majority of their public-sector services through online channels. This meant that 188 European country ministries and 35,000 local authorities were potential buyers of Alpha's services. Alpha had developed expertise in this market since 1998, but the majority of Alpha's international competitors in that particular area did not start their build-up of expertise until after 2000.

The investment of Alpha in e-government services was followed by the formation of Alpha International Ltd in 2002. This move aimed to facilitate Alpha's penetration of the private and public sector of the EU in order to expand its customer base and avoid dependency on its big customers. This practice proved to be prudent a few years later, when significant contracts were signed with the European Commission. Two subsidiaries, in Brussels and Dublin, supported the company's interests in the European institutions and performed duties such as sales, pre-sales and sales promotion with other European public administration customers. Alpha essentially became a micromultinational (see Dimitratos *et al.*, 2003), and its foreign direct investment facilitated the exploitation of opportunities abroad.

Investment in language-related services – the search for financing

Unfortunately, in the summer of 2002, the earlier concerns of the company's management team were realized. As the recession hit the Greek economy, and in particular the ICT sector (whose growth had decreased at an annual rate of 3.4 per cent in the previous year), Microsoft (from which more than half of the company's turnover was accrued at that time), decided to reduce significantly its outsourcing orders for localization. For the first time since the management buy-out, G.K. and G.S. began to feel that the future growth of the company was threatened. Sales projections for the following year showed a decrease of 10 per cent, causing concern in an industrial sector challenged by 'anxiety'. Although this was big shock for the company, Alpha was able to confront it. The management team, always alert, was evaluating new ways of helping Alpha to sustain its position and encourage further growth. An opportunity appeared in the translation market. Specifically, the enlargement of the EU with the addition of ten new member countries was about to create a 'Tower of Babel' effect within the boundaries of Europe, making translation services profitable. Alpha had long-established capabilities in language-related services through offering integrated globalization services for software and websites. Apart from its own skills, the company had formed an important network of language partners all over Europe. Hence, this was a great challenge, but also a tough decision to make. Could Alpha, a high-tech IT company, also accept a role as a traditional translation service provider? Despite Alpha having the technical capabilities to exploit the new opportunity, did it have the financial resources to support a new investment? Whether risky or perspicacious, G.K and G.S. decided to go ahead.

The persistent recession in the IT sector resulted in stagnant orders from Alpha's big international customers, who were to turn to lower-cost suppliers worldwide to cover the greater part of their localization needs. The investment made in e-government services had not yet produced positively significant results by 2003 and, to make matters even worse, the projects driven by the Third Community and Support Framework concerning Information Society tenders in Greece were showing significant delays in their implementation by the Greek government.

The end of 2003 was the toughest period in the company's history, and serious cash-flow problems were anticipated for early 2004. The company's need for external financing was imperative in order to preserve its investments and support its growth. G.K. and G.S., equipped with a well-written business plan, launched a quest for financing in the beginning of 2004. Taking on the risk of their decisions and providing personal guarantees, they opted for and succeeded in gaining a long-term bank loan of approximately €1.5 million. This evidence suggests that the personalities and traits of the top management team of the entrepreneurial firm are strongly imprinted on the company, a fact widely encountered in entrepreneurship literature. The investment in the language sector turned out to offer profitable opportunities for growth to the international new venture.

The most recent situation

At the time of writing, Alpha has a permanent workforce of seventy highly specialized staff committed to the mission and the quality engagement of the company with its customers. The relationship between management and employees is more personal than a distant employer–employee relationship, and thus high employee retention rates have been preserved. According to G.K., 'The company owes its success to its people.' Well-balanced education and young age profiles are maintained; 85 per cent of the employees are involved in the production and technical department and the remaining 15 per cent in administration and sales. The headquarters of the company is in Athens in a $2200\,m^2$ office space from where local market sales, as well as sales in the markets of Eastern Europe, the Balkans and the ten new members of the enlarged EU are co-ordinated. Alpha International Ltd abroad supports the company's interests in other European countries.

Alpha is committed to maintaining *its infrastructure at the top of the most demanding industrial requirements*; and it has established synergies and alliances with some of the biggest corporations, both internationally and within Greece, and with major European research institutions and universities to further support its 'big account mentality' customer service and satisfaction. Network formations have typically been shown to assist the small firm in overcoming resource deficiencies (Aldrich and Martinez, 2001; Aldrich and Zimmer, 1986) and internationalize effectively (Crick and Jones, 2000; Westhead *et al.*, 2001), and thus have constituted the source of knowledge and resource creation of the firm.

The road ahead

Alpha has always benefited from a profitable balance sheet across the years. Even when costly investments created cash flow problems for the company, it achieved sustainable growth, albeit occasionally slow. The same is true for profitability. The turnover for the fiscal year 2003 reached €5.7 million with net profits of €0.5 million. Having secured income from long-term contracts, growth prospects seem very promising, potentially even reaching the rates of the company's hyper-growth phase.

The international new venture's strengths have always been its excellent technical expertise and its focus on customer satisfaction. The highly specialized staff, international partner network, European presence, competitive pricing, 'big account' mentality and culture, and continuous investment in R&D and training are some of the numerous qualities that distinguish Alpha, and have provided a strong base for the company's success. In line with the resource-based view (Barney, 1986; Penrose, 1959; Wernerfelt, 1984), the firm has successfully nurtured those competencies likely to result in income from the international (as well as Greek) marketplace. For the future, Alpha is already well positioned in the market to offer integrated content services, as the company foresaw the potential demand for these, mainly within the European Union, well before other international companies were aware of it. The management team of Alpha is convinced that integrated content services in the EU offers a major business opportunity to grasp.

Thus Alpha's vision seems clear regarding the road ahead: 'to serve the tremendously important market need for integrated content services, which will be the trend in the years to come'. It appears that this is an area where profitable opportunities will lie for Alpha. Nevertheless, the owners of the company, being guided by the company's history, are not overconfident. They know that the entrepreneurial process is dynamic and not always predictable. The road ahead is promising but they must apply continuous effort and make investments. The growth of the firm shows that its progression is based on its prior successes, mistakes and failures, stressing a path-dependence growth, in agreement with the resource-based view. Learning has taken place for Alpha, which has aided its route to growth, as the entrepreneurship literature suggests (Gaglio, 1997; Minniti and Bygrave, 2001).

Concluding remarks

Alpha has always relied on its strong technical expertise and its focus on customer satisfaction. Its highly specialized staff, international partner network, tight co-operation and success with 'big accounts', pan-European presence, competitive pricing schemes, continuous investments in R&D and employee training constitute the foundation of the firm's growth. These elements form the core competencies of the firm in its pursuit of profitable market opportunities. The evidence from this research suggests that these core competencies are developed and nurtured in numerous departments and functions of the firm, such as marketing, R&D and human resources, stressing that key resources and capabilities require a company-wide holistic effort in order to be cultivated and subsequently assist in the exploitation of opportunities by small firms.

In addition, the case study evidence suggests that the success of the firm is largely based on a 'path-dependent' stock of knowledge, whereby this knowledge has been gained in the quest for profitable market opportunities. The company effectively switched between (mainly international) geographic markets as well as between industrial product markets (ICT, e-government, translation), forming a portfolio of intertwined strategic choices in which growth was sought and achieved. Thus the results of this research show that the growth of the small firm is a comprehensive phenomenon not likely to be constrained within the limits of any geographical or product market. On the contrary, it appears that activities of successful entrepreneurs seem to transcend geographic markets and industry sectors, wherever opportunities may emerge. In doing so, the this international new venture effectively managed to avoid considerable risks abroad, such as cross-cultural and commercial risks stemming from the liability of being foreign, and significant losses of orders from Microsoft and IBM, respectively (see Cavusgil, 2006). Alpha has managed to survive and grow in a period of anxiety, where the whole ICT sector faced significant problems. Viewed in this light, this case study has also offered some evidence on the theme of the crucial role of internationalization in the early growth of small firms, and the way that strategic decisions in product markets interact with the internationalization process (Bell *et al.*, 2004).

Future research may explore to a greater extent the role of dynamic learning in the internationalized small firm across different markets. It appears that firms and entrepreneurs learn from their mistakes and failures (Gaglio, 1997), and it would be useful to explore the constituents and mechanisms of such useful knowledge creation in the learning process. Related to this issue is the topic of whether some of the knowledge gained is of a rather ineffectual type to the entrepreneur. If this holds, how can entrepreneurs avoid this ineffectual knowledge, if at all? Also, the evidence of this research accentuates the fact that growth of the firm takes place wherever opportunities emerge in international or domestic marketplaces and in various product markets. Hence it seems that the examination of either international or domestic activities alone offers a one-sided view of the complex nature of strategic growth and exploitation of opportunities for the firm. Researchers in strategic management and international business would do better to examine the theme of small firm growth holistically in order to gain a comprehensive picture of this theme. In a similar vein, international entrepreneurship study would benefit if it incorporated notions from the (mainstream) entrepreneurship literature in relation to exploitation of opportunities and the growth of the small firm. Viewed in this light, a blend of the strategic choice and firm growth theories from strategic management, international business and entrepreneurship literature streams can illuminate aspects of the internationalized small firm growth phenomenon.

With regard to the implications of the study for management practitioners, it seems that the growth of a small firm takes place over a long period, in which both favourable and adverse events can affect the company's performance. Effective entrepreneurs appear always to 'keep an eye' on potentially profitable prospects as well as to learn from their failures and insist on the aim of growth for their firm. 'Keep walking' is likely to be the best lesson for small firm managers from this current study, since successful entrepreneurs tend to pursue steadfastly the objective of growth, gain experience and knowledge, avoid mistakes and attempt to alleviate the consequences of negative events.

References

Aldrich, H. E. and M. A. Martinez (2001) 'Many Are Called, but Few Are Chosen: An Evolutionary Perspective for the Study of Entrepreneurship', *Entrepreneurship Theory and Practice*, 25(4), 41–56.

Aldrich, H. and C. Zimmer (1986) 'Entrepreneurship through Social Networks', in D. L. Sexton and R. W. Smilor (eds), *The Art and Science of Entrepreneurship* (Cambridge, Mass.: Ballinger), 3–23.

Ansoff, I. H. (1965) *Corporate Strategy: An Analytic Approach to Business Policy for Growth and Expansion* (New York: McGraw-Hill).

Autio, E., H. J. Sapienza and H. J. Almeida (2000) 'Effects of Age at Entry, Knowledge Intensity, and Imitability on International Growth', *Academy of Management Journal*, 43, 909–24.

Barney, J. B. (1986) 'Strategic Factor Markets: Expectations, Luck and Business Strategy', *Management Science*, 32, 1231–41.

Bell, J., D. Crick and S. Young (2004) 'Small Firm Internationalization and Business Strategy', *International Small Business Journal*, 22(1), 23–56.

Cavusgil, S. T. (2006) 'International Business in the Age of Anxiety: Company Risks', Opening presentation at the 33rd Academy of International Business (UK Chapter) Conference, Manchester, 7–8 April.

Crick, D. and M. V. Jones (2000) 'Small High-technology Firms and International High-technology Markets', *Journal of International Marketing*, 8(2), 63–85.

Crick, D. and M. Spence (2005) 'The Internationalisation of "High Performing" UK High-tech SMEs: A Study of Planned and Unplanned Strategies', *International Business Review*, 14(2), 167–85.

Dimitratos, P. and M. V. Jones (2005) 'Future Directions for International Entrepreneurship Research', *International Business Review*, 14(2), 119–28.

Dimitratos, P. and E. Plakoyiannaki (2003) 'Theoretical Foundations of an International Entrepreneurial Culture', *Journal of International Entrepreneurship*, 1, 187–215.

Dimitratos, P., J. E. Johnson, J. Slow and S. Young (2003) 'Micromultinationals: New Types of Firms for the Global Competitive Landscape', *European Management Journal*, 21(2), 164–74.

Gaglio, C. M. (1997) 'Opportunity Identification: Review, Critique and Suggested Research Directions', in J. A. Katz (ed.), *Advances in Entrepreneurship, Firm Emergence and Growth*, vol. 3 (Greenwich, Conn.: JAI Press), 139–202.

Johanson, J. and J.-E. Vahlne (1977) 'The Internationalization Process of the Firm: A Model of Knowledge Development and Increasing Foreign Market Commitments', *Journal of International Business Studies*, 8(1), 23–32.

Jones, M. V. (1999) 'The Internationalization of Small High-technology Firms', *Journal of International Marketing*, 7(4), 15–41.

Jones, M. V. (2001) 'First Steps in Internationalisation – Concepts and Evidence from a Sample of Small High-technology Firms', *Journal of International Management*, 7, 191–210.

Katsikeas, C. S., S. L. Deng and L. H. Wortzel (1997) 'Perceived Export Success Factors of Small and Medium-sized Canadian Firms', *Journal of International Marketing*, 5(4), 53–72.

Kuemmerle, W. (2002) 'Home Base and Knowledge Management in International Ventures', *Journal of Business Venturing*, 17(2), 99–122.

Leonidou, L. C., C. S. Katsikeas and S. Samiee (2002) 'Marketing Strategy Determinants of Export Performance: A Meta-analysis', *Journal of Business Research*, 55, 51–67.

Madsen, T. K. and P. Servais (1997) 'The Internationalization of Born Globals: An Evolutionary Process?', *International Business Review*, 6, 561–83.

Makridakis, S., Y. Caloghirou, L. Papagiannakis and P. Trivellas (1997) 'The Dualism of Greek Firms and Management: Present State and Future Implications', *European Management Journal*, 15, 381–402.

McDougall, P. P. (1989) 'International versus Domestic Entrepreneurship: New Venture Strategic Behavior and Industry Structure', *Journal of Business Venturing*, 4, 387–400.

McDougall, P. P., S. Shane and B. M. Oviatt (1994) 'Explaining the Formation of International New Ventures: The Limits of Theories from International Business Research', *Journal of Business Venturing*, 9, 469–87.

Minniti, M. (2004) 'Entrepreneurial Alertness and Asymmetric Information in a Spin-glass Model', *Journal of Business Venturing*, 19, 637–58.

Minniti, M. and W. Bygrave (2001) 'A Dynamic Model of Entrepreneurial Learning', *Entrepreneurship Theory and Practice*, 25(3), 5–16.

OECD (2002) *OECD Information Technology Outlook: ICTs and the Information Economy* (Paris: OECD Publication Services).

Oviatt, B. M. and P. P. McDougall (1994) 'Toward a Theory of International New Ventures', *Journal of International Business Studies*, 25, 45–64.

Oviatt, B. M. and P. P. McDougall (1995) 'Global Start-ups: Entrepreneurs on a Worldwide Stage', *Academy of Management Executive*, 9(2), 30–43.

Oviatt, B. M. and P. P. McDougall (2005) 'Defining International Entrepreneurship and Modelling the Speed of Internationalization', *Entrepreneurship Theory and Practice*, 30, 537–53.

Park, S. and Z.-T. Bae (2004) 'New Venture Strategies in a Developing Country: Identifying a Typology and Examining Growth Patterns through Case Studies', *Journal of Business Venturing*, 19, 81–105.

Patton, M. Q. (2002) *Qualitative Evaluation and Research Methods*, 3rd edn (Newbury Park, Calif.: Sage).

Penrose, E. (1959) *The Theory of the Growth of the Firm* (London: Blackwell).

Prahalad, C. K. and G. Hamel (1990) 'The Core Competence of the Corporation', *Harvard Business Review*, 68(2), 79–91.

Shapero, A. (1982) 'Social Dimensions of Entrepreneurship', in C. A. Kent, D. L. Sexton and K. H. Vesper (eds), *The Encyclopedia of Entrepreneurship* (Englewood Cliffs, NJ: Prentice-Hall), 72–90.

Shook, C. L., R. L. Priem and J. E. McGee (2003) 'Venture Creation and the Enterprise Individual: A Review and Synthesis', *Journal of Management*, 29, 379–400.

Voudouris, I., S. Lioukas, S. Makridakis and Y. Spanos (2000) 'Greek Hidden Champions: Lessons from Small, Little-known Firms in Greece', *European Management Journal*, 18, 663–74.

Wernerfelt, B. (1984) 'A Resource-based View of the Firm', *Strategic Management Journal*, 5, 171–80.

Westhead, P., M. Wright and D. Ucbasaran (2001) 'The Internationalization of New and Small Firms: A Resource-based View', *Journal of Business Venturing*, 16, 333–58.

Yin, R. K. (1989) *Case Study Research – Design and Methods* (Newbury Park, Calif.: Sage).

Zahra, S. A., D. R. Ireland and M. A. Hitt (2000) 'International Expansion by New Venture Firms: International Diversity, Mode of Market Entry, Technological Learning, and Performance', *Academy of Management Journal*, 43, 925–50.

12

Cultural Adaptation in Cross-border Web Presence: An Investigation of German Companies' Domestic, US, UK and Latin American Websites

Matthias Hossinger, Rudolf R. Sinkovics and Mo Yamin

Introduction

At the time of writing, nearly a billion people are connected to the Internet (Okazaki, 2004). Between 2000 and 2005 the Internet experienced a growth of 160 per cent (Internet World Stats, 2005), making it a source of about USD 3.2 trillion in revenues for businesses and their e-commerce activities (Singh *et al.*, 2003). Hence, web-presence is arguably crucial to business-success (Alvarez *et al.*, 1998). However, while websites are virtually accessible to anybody from anywhere, truly tapping into online customers involves more than simply putting up a website. Website design and content arguably needs to take into account linguistic and cultural differences (Singh *et al.*, 2003). As Lim *et al.* have observed, the virtual space on the Internet is not boundary-less or culture-free (Lim *et al.*, 2004). Similar observations have been made by, for example, Singh and Baack (2004) and Singh *et al.* (2004).

Paralleling the long-standing debate on standardization versus adaptation in international marketing (see, for example, Agrawal, 1995; Theodosiou and Leonidou, 2003), there is discussion on whether, in the online domain, adaptation to local cultural manifestations or standardization to effectively transmit online content is the more appropriate strategy. There is also a lack of empirical evidence in the fields of 'applied culture' on the Internet; hence perspectives on whether a 'contingency perspective' is appropriate, and in which context, are limited. A further problem is that most of the extant literature focuses heavily on US culture and companies (Okazaki, 2004; Singh *et al.*, 2003). Moreover, we witness methodological weaknesses such as relatively small sample sizes (Fink and Laupase, 2000; Singh and Baack, 2004; Singh *et al.*,

2004; Singh *et al.*, 2003) and a somewhat limited depth of cultural analysis (Okazaki and Rivas, 2002).

Following a review of the literature, this chapter replicates and extends the methodological approach suggested by Singh *et al.* (2005a). We explore whether MNCs and internationally operating German firms favour standardized or adapted online communication strategies, both in their own domestic habitat and in country-specific websites in the USA, UK and in Latin America. Cultural values are measured using Hofstede's (1991) and Hall's (1976) dimensional approach.

Conceptual background

The standardiztion versus adaptation debate

The standardization/adaptation debate in international marketing has inspired academics and practitioners for more than three decades. Arguments of cost reduction, scale- and scope-effects, brand building as well as meeting customer demands, and culture-bound preferences and expectations have resulted in contributions on either the standardization or adaptation side of the continuum (Alashban *et al.*, 2002; Buzzell, 1968; Fatt, 1967; Levitt, 1983; Mueller, 1992; Papavassiliou and Stathakopoulos, 1997; Rutigliano, 1986; Walters, 1986; Yip, 1989).

The standardization approach has been criticized by many practitioners and marketing professionals as this approach is likely to be too product-orientated. As Douglas and Wind (1987) point out, standardization *per se* 'implies a product orientation, and a product driven strategy, rather than a strategy grounded in the systematic analysis of customer behaviour and response patterns and market characteristics'. (p. 19). Negative implications of pure product-orientated strategies have also been outlined by Levitt (1960) and Laughlin *et al.* (1994). Cavusgil and Zou (1994) also point to the disadvantages in terms of vulnerability to competitive attacks (see also Ricks, 1999; Zou *et al.*, 1997) and others point at the lack of responsiveness to diverse governmental, economical/ ecological and socio-cultural settings (Doz and Prahalad, 1980; Zou and Cavusgil, 1996).

The contingency perspective (see, for example, Agrawal, 1995; Cavusgil *et al.*, 1993) has removed the binary choice element from the discussion, suggesting that the decision will depend on issues such as product category, industry, competition (see, for example, Jain, 1989; Kustin, 2004; Quelch and Hoff, 1986; Subramaniam and Hewett, 2004; Theodosiou and Leonidou, 2003; Walters and Toyne, 1989). This adaptive strategy has been examined empirically by numerous authors with various degrees of empirical sophistication (see, for example, Agrawal, 1995; Green *et al.*, 1975; Johansson, 1994; Mueller, 1992; Onkvisit and Shaw, 1987). Katsikeas *et al.*

Theodosiou (2006) provide what is in our view the most rigorous empirical examination of this perspective.

Taking the issue online

Given the very large and rapidly growing number of Internet users in many countries there is an enormous growth potential for online commerce in both the b-to-b and b-to-c markets. Attractions exist for managers in terms of cost savings in the online domain (Quelch and Klein, 1996) and efficiency effects of market transactions (Petersen *et al.*, 2002). These arguments implicitly suggest a standardized approach to online communication.

Yet information technology competences and capabilities of companies, such as operating websites or conducting business online, can easily be replicated by competitors (Evans and Wurster, 1997; Yamin and Sinkovics, 2006). Competitive advantages on the Internet are therefore not likely to be sustainable (see, for example, Carr, 2003; Riquelme, 2002) and companies must seek differentiation advantages. Kotha (1998) suggests that uniform communication patterns may not be sufficient to maintain healthy profit margins and a competitive advantage. Yip and Dempster (2005) concur with Porter, and suggest the establishment of a 'unique set of activities'; they argue that companies in global industries must 'carefully monitor how rivals are making use of the Internet, and lead or match rivals' activities' (Yip and Dempster 2005, p. 5). In all, the debate over online adaptation or standardization has not yet provided a consolidated view on the issue.

Standardization on the Internet

For Singh and Boughton (2002), a standardized website entails 'the same web content, in the same language, for both domestic and international users. Websites under this category, do not prominently display any information about their international operations' (Singh and Boughton, 2002, p. 302). It is argued that online standardization leads to cost savings (see, for example, Kambil, 1995; Sinkovics and Penz, 2005) as well as the strengthening of the global brand (see Yip and Dempster, 2005). Tsikriktsis (2002) concludes that culture plays a 'significantly less important role in Web site quality expectations compared with traditional service quality expectations' (see Tsikriktsis, 2002, p.110). Similarly, Forrester Research posits that repeat visits to websites are predominantly determined by interactivity, trust, the right composition of quality content, ease of use, speed and frequency of updating; and cultural dimensions and appeal are of negligible importance.

Overall, proponents of online standardization claim that it is the preferred strategy for 'pushing' visitors through the conversion process from being 'surfers' to 'purchasers' (Berthon *et al.*, 1996). On the other hand, proponents of adaptation strategy argue that standardization does not generate distinctiveness in web communication, and hence cannot maximize market potential in respective markets, and thus risks losing competitive advantage.

Adaptation on the Internet

According to Singh and Boughton (2002), adapted websites exhibit specific time, date, zip code and number formats. These sites have country-specific templates reflected in the country-specific unique resource locators (URLs) such as, .de (Germany), .com (USA), and .co.uk (UK). Furthermore, these country-specific sites are featured visibly on the parent company websites and pay detailed attention to cultural differences, most notably language issues.

Discomfort with the standardization of online content was expressed by, for example, Hodges (1997), who noted that English was an inappropriate language with which to target Japanese internet shoppers. Academics such as Yip (2000), Lynch *et al.* (2001), Fink and Laupase (2000), Marcus and Gould (2000), Zahir *et al.* (2002) or Cyr and Trevor-Smith (2004) reveal that culture does influence the design of Internet sites. Hence, in practical terms, cultural divergence is happening. Looking at the profiles of Internet buyers in twenty countries, it has been suggested that "even with increased electronic interactions, people still need to feel engaged (culturally or contextually) with vendors, even online. Consequently, companies that have an understanding of and an ability to 'mirror' the culture of their target country will have a competitive global advantage" (Lynch and Beck, 2001, p. 735). Samiee's (1998) suggestion that culture can have a major impact on the success of e-marketing efforts appears to be valid also in online contexts. Empirical evidence is provided by Luna *et al.* (2002), who show that exposure to culturally appropriate websites reduces the cognitive efforts required from customers. The enhanced quality and frequency of online contact is likely to result in conversion efficiency. Chakraborty *et al.* (2005) provide empirical evidence from a b-to-b context, which suggests that 'one size fits all' approaches in website design should be avoided, and websites should be custom-built for geographical regions.

Cultural analysis on the web

Singh (2002) argues that Hofstede's (1984, 1991) cultural dimensions represent a valuable framework for research on web analysis, advertising and web content development. Using Hofstede's (1984) four cultural dimensions and Hall's (1976) context dimensions, Singh *et al.* (2003) evaluated the level of cultural adaptation of American companies' domestic and Chinese websites. They state, 'the web is not a culturally neutral medium, but it is full of cultural markers that give country-specific websites a look and feel unique to the local culture' (Singh *et al.*, 2003, p. 63).

Building on the same cultural framework, Singh and Baack (2004) compared US and Mexican websites for differences in the depiction of local cultural values online. Mexican websites demonstrated higher scores in collectivism and power distance as well as slightly higher ones on masculinity. Interestingly, uncertainty avoidance was not in line with Hofstede's results. Further studies illustrated that the content of US and Japanese websites reflected national cultural values (Singh and Baack, 2004) and similar results were encountered in investigations of Chinese, Indian, Japanese and US websites (Singh, Zhao and Hu, 2005). 'The results indicate that local websites of India, China, Japan and U.S. not only reflect cultural values of the country of their origin, but also seem to differ significantly from each other on cultural dimensions' (Singh, Zhao and Hu, 2005, p. 138). Similar work, comparing b-to-c e-commerce firms in the USA, France and Germany 'show[s] that the cultural values presented in the local websites of the three countries are linked to theoretical work regarding the cultural differences between the countries' (Singh, Kumar and Baack, 2005, p. 82). Within this study, not only country-values are compared, but also US headquarters websites targeting the French and German country markets. Overall, these findings strengthen the robustness of Hofstede's work over time, and the validity of the content analysis framework used.

Hypotheses development

The six cultural dimensions used in this study were first operationalized by Singh *et al.* (2003) later revised and updated (Singh, Zhao and Hu, 2005) to comprise six dimensions and twenty three cultural categories (see Table 12.1).

Table 12.1   Cultural value framework

Collectivism
a_1 *Community relations* – presence or absence of community policy, giving back to community, social responsibility policy
a_2 *Clubs or chat rooms* – presence or absence of members' club, product-based clubs, chats with company employees, chat with interest groups, message boards, discussion groups and live talks
a_3 *Family theme* – pictures of family, pictures of teams of employees, mention of employee teams, emphasis on team and collective work responsibility in vision statement or elsewhere on the website, emphasis on customers as a family
a_4 *Loyalty programmes* – frequent flyer programmes, customer loyalty programmes, and company credit cards for local country, special membership programmes
a_5 *Newsletter* – online subscriptions, magazines and newsletters

Uncertainty avoidance
b_1 *Customer service* – FAQs, customer service option, customer help or customer service e-mails
b_2 *Guided navigation* – site maps, well-displayed links, links in the form of pictures or buttons, forward, backward, up and down navigation buttons
b_3 *Local stores* – mention of contact information for local offices, dealers and shops

b_4 *Local terminology* – use of country-specific metaphors, names of festivals, puns, a general local touch in the vocabulary of the web page

b_5 *Tradition theme* – emphasis on history and the ties of a particular company with a nation, emphasis on respect, veneration of one's elders, phrases such as 'most respected company', 'keeping the tradition alive', 'for generations', 'company legacy', etc.

Power distance

c_1 *Company hierarchy information* – information about the ranks of company personnel, information on organizational charts, information about country managers

c_2 *Pictures of CEOs* – pictures of executives, important people in the industry or celebrities

c_3 *Proper titles* – titles of the important people in the company, titles of the people in the contact information, and titles of people in the organizational charts

c_4 *Vision statement* – statement by the CEO or company head about the vision of the company

Individualism

d_1 *Independence theme* – images and themes depicting self-reliance, self-recognition, achievement

d_2 *Product uniqueness* – unique selling points of the product, product differentiation features

d_3 *Personalization* – features such as gift recommendations, individual acknowledgements or greetings, web page personalization

High context

e_1 *Aesthetics* – attention to aesthetic details, liberal use of colours, bold colours, emphasis on images and context, use of love and harmony appeal

e_2 *Politeness and indirectness* – greetings from the company, images and pictures reflecting politeness, flowery language, use of indirect expressions such as 'perhaps', 'probably' and 'somewhat', overall humbleness in company philosophy and corporate information

e_3 *Soft sell approach* – use of affective and subjective impressions of intangible aspects of a product or service, high usage of entertainment themes to promote the product

Low context

f_1 *Hard sell approach* – discounts, promotions, emphasis on product advantages using explicit comparison

f_2 *Rank or prestige of the company* – features such as company rank in the industry, listing in *Forbes* or *Fortune*, numbers showing the growth and importance of the company

f_3 *Use of superlatives* – use of superlative words and sentences such as 'the number one...', 'the top company', 'the leader', 'world's largest', etc.

Source: Singh *et al.*, 2005, pp. 77, 79–80.

Following the literature review above, we suggest that adapted websites will depict cultural values online, which are representative of respective country values (as provided by Hofstede and Hall). Detailed hypotheses will be provided below. The overall hypothesis is that:

> **H1 Cultural value depiction will differ between German domestic websites and their international subsidiary websites. Cultural values depicted online will correspond to the respective target market cultural values (that is US, UK and Latin American cultural values).**

Individualism–collectivism

Individualistic societies are more 'I-conscious', reflecting traits such as self-reliance, independence, achievement and freedom (Gudykunst, 1998; Hofstede, 1984). The reverse is true for collectivist societies. Taking into consideration all the nations analysed for this chapter, the US scores highest on individualism, with 91 points out of a 100, followed by the UK (89) and Germany (67) (Hofstede, 1991). Latin American countries score lower on individualism, but higher on collectivism. We therefore conclude:

> **H2a German companies' home country websites will depict lower levels of individualistic elements than their US websites.**

> **H2b German companies' home country websites will depict lower levels of individualistic elements than their UK websites.**

> **H2c German companies' home country websites will depict higher levels of individualistic elements than their Latin American websites.**

While in Hofstede's work, individualism and collectivism form a bipolar axis, Singh *et al.* (2005b) use these as independent measures. The argument is that 'individual and collectivist tendencies can co-exist in societies, and . . . it is more appropriate to treat them as separate dimensions' as the work of Triandis (1994), Han and Shavitt (1994) and Cho *et al.* (1999) implies. Since Latin America – on average – depicts the highest level of collectivism, the following hypotheses can be derived:

> **H3a German companies' Latin American websites will depict higher levels of collectivistic elements than their US websites.**

> **H3b German companies' Latin American websites will depict higher levels of collectivistic elements than their UK websites.**

> **H3c German companies' Latin American websites will depict higher levels of collectivistic elements than their German websites.**

Uncertainty avoidance

The Internet, as a new medium, brings about a certain degree of uncertainty, hence 'people from high uncertainty avoidance (UA) cultures need more reassurance and uncertainty reduction features to facilitate their online purchases' (Singh, Zhao and Hu, 2005, p. 134). It is therefore suggested that countries with high UA scores, such as Germany (65) and Latin America (86), will prefer sites with limited, 'simple, and redundant navigational devices'. Individuals within the USA (46) and the UK (35), on the other hand, 'would tend to prefer greater complexity and less control over navigation', as 'low UA cultures tend to desire more informal business arrangements and are more relaxed' (Bernard, 2003). This leads to the fourth set of hypotheses:

H4a   German companies' home country websites will depict higher levels of uncertainty avoidance elements than their US websites.

H4b   German companies' home country websites will depict higher levels of uncertainty avoidance elements than their UK websites.

H4c   German companies' home country websites will depict lower levels of uncertainty avoidance elements than their Latin websites.

Power distance

Latin American countries score very high on power distance (PD), with an average score of 70. There is appreciation of social status, referent power, authority and legitimacy (Singh, Zhao and Hu, 2005). The USA (40) scores low on PD, with Germany and the UK scoring even lower at 35. Hence gender-neutral websites that de-emphasize hierarchical differences between individuals are encouraged (Bernard, 2003). Building on work that relates high on PD appeals to high PD societies (Singh, Zhao and Hu, 2005; Zandpour *et al.*, 1994), the subsequent list of hypotheses is brought forward:

H5a   German companies' home country websites will depict slightly lower levels of power distance elements than their US websites.

H5b   German companies' home country websites will depict the same levels of power distance elements as their UK websites.

H5c   German companies' home country websites will depict lower levels of power distance elements than their Latin websites.

High- and low-context cultures

Communication in high-context cultures is implicit, indirect and deeply embedded in the context, while in low-context cultures it is more direct, less implicit and more informative (Singh, Zhao and Hu, 2005). According to Hall (1976), Germany, the USA and the UK are low-context cultures, whereas Latin America is more orientated to highly contextualized signals. We thus put forward the following hypotheses:

H6a   **German companies' home country websites, their US and UK websites will depict higher levels of low-context elements than their country-specific Latin American websites.**

H6b   **Country-specific Latin American websites of German companies will depict the highest level of high-context elements compared to their home country websites, country-specific UK and US websites.**

Empirical study

Research context

From the outset, we decided to involve a German, English or Spanish-speaking selection of countries in the study. This was a function of cultural diversity as well as a pragmatic choice, given the language capabilities of the research team. Furthermore, we used the World Economic Forum's Network Readiness Index (NRI), to support our choice of countries. The NRI measures the propensity for countries to exploit the opportunities offered by information and communications technology and is published annually (World Economic Forum, 2005). We chose German companies as the unit of analysis, with export destinations of the US, UK and Latin American markets. We merged Latin American countries into one group because, while German companies are present in many of the Latin American markets, only a fraction of the companies have establishments in more than one Latin American market (see Table 12.2).

Research design

Information on German companies, exporting to the UK, USA and Latin American markets was collected from the AMADEUS database, a pan-European database containing information on 7 million publicly and privately listed companies in thirty-eight countries. A random sample of companies was chosen for website analysis.

Exporting German companies were only considered for inclusion in the sample if foreign company or institutional investor holdings were more than 49 per cent. A total of 100 German exporting companies, mainly MNCs, which had an Internet presence in Germany, the USA and the UK were included in the study (that is, 300 websites in total). And websites from these companies located in Latin American markets were included in the analysis. On average,

Table 12.2 Market characteristics and NRI rankings for Latin American countries

	PPP GDP in USD million	PPP GDP per capita in USD	NRI rankings				†	Internet users	Usage growth	Percentage population (penetration)
			01/02[a]	02/03[b]	03/04[c]	04/05[d]				
Argentina	486,366	12,400	32th	45th	50th	76th	56th	7,500,000	200.0	20.0
Chile	183,286	10,700	34th	35th	32nd	35th	58th	4,000,000	127.6	25.8
Colombia	322,582	6,600	57th	59th	60th	66th	67th	3,585,688	308.4	7.8
Costa Rica	39,823	9,600	45th	49th	49th	61th	91th	1,200,000	380.0	27.9
Ecuador	51,330	3,700	71th	75th	89th	95th	90th	581,600	223.1	4.8
Mexico	1,014,514	9,600	44th	47th	44th	60th	26th	14,901,687	449.4	14.3
Peru	155,388	5,600	52th	67th	70th	90th	87th	4,570,000	82.8	16.3
Guatemala	52,926	4,200	68th	73th	86th	88th	99th	400,000	515.4	3.2
Panama	22,235	6,900	48th	61th	58th	69th	103th	192,100	326.9	6.2
Salvador	34,396	4,900	55th	63th	62th	70th	106th	550,000	1275.0	8.5
Uruguay	32,174	14,500	37th	55th	54th	64th	114th	1,190,120	221.7	36.6
Venezuela	155,790	5,800	50th	66th	72th	84th	74th	3,040,000	220.0	12.2
Latin America	2,550,810							41,711,195	360.9	15.3

Notes and sources: [a] Out of 75 countries; [b] Out of 82 countries (Dutta and Jain, 2004); [c] Out of 102 countries; [d] Out of 104 countries (Lopez-Claros and Dutta, 2005). † Export destination rank for Germany: Out of 228 trading partners (Federal Statistical Office Germany, 2004). The figures for Latin America are calculated as follows: (i) SUM for PPP GDP in USD millions, SUM for internet users; (ii) MEAN aggregate for usage growth; (iii) MEAN aggregate for percentage population (penetration). * Latin American follows: usage growth and percentage population penetration represents changes are all Latin American countries.

217

we spent about 20 minutes assessing each website by means of a questionnaire-type assessment-sheet.

Research instrument

The research instrument for the evaluation of the websites comprised the questionnaire-type assessment-sheet used by two independent researchers to rate the cultural value depiction of the websites. The six cultural dimensions for this study are listed in Table 12.1 above. Measures for each of the cultural dimensions – operationalized on a five-point Likert scale – ranged from 'not depicted' to 'prominently depicted'.

Inter-coder reliability for the two raters was very satisfactory, at 92.3 per cent. Cronbach's alpha coefficient also produced very good results, with scale reliabilities for each of the six dimensions greater than 0.8. Overall, for each of the six dimensions, we found that the countries' websites featured significantly different results (Wilks' Lambda=0.634; p=0.000). Detailed results are presented below.

Data analysis and findings

The composite score on all six cultural dimensions shows that German domestic, US, UK and Latin American websites differ significantly from each other (means: Germany=3.08, USA=2.98, UK=2.95, Latin America=2.15; F=159.28; p<0.05). Apart from high context countries, all the other cultural dimensions differed significantly across the local websites of Germany, the USA, the UK and Latin America, while the Tukey (HSD) *post-hoc* test revealed that there was no statistical difference between US and UK sites. Overall, H1 was supported see (Table 12.3).

Individualism. According to a MANOVA test, there are significant differences among the four independent variables (means: Germany=3.30, USA=3.28, UK=3.24, Latin America=2.17; F=23.18; p<0.05). However, Tukey's (HSD) *post-hoc* test H2a cannot be supported, neither can H2b. However, German websites show higher individualism than Latin American sites, that is, H2c is supported.

Collectivism. No clear picture emerged for the collectivism dimension. While we found significant differences overall regarding collectivism, results contradicted hypothesized relationships and Hofstede findings. Latin American websites demonstrated less collectivistic attributes compared to their German domestic, US and UK websites (no support for H3a, H3b or H3c).

Uncertainty avoidance. German and country-specific US and UK websites do not show statistically significant differences in terms of uncertainty avoidance (H4a, H4b are not supported). However, uncertainty avoidance featured more

Table 12.3 Descriptive statistics, MANOVA and Tukey (HSD) *post hoc* results

Dimensions	German company websites in									
	Germany (n = 100)		USA (n = 100)		UK (n = 100)		Latin America (n = 47)		F-value	Sig.
	Mean	SD	Mean	SD	Mean	SD	Mean	SD		
Collectivism	2.7580^a	0.8919	2.5540^a	0.8149	2.5720^a	0.8135	1.9783^b	0.6841	9.73*	0.000
Individualism	3.3003^a	0.8338	3.2771^a	0.8701	3.2438^a	0.8516	2.1711^b	0.7931	23.18*	0.000
Uncertainty avoidance	3.6020^a	0.5891	3.4780^a	0.5804	3.4660^a	0.5784	2.9383^b	0.5642	14.52*	0.000
Power distance	3.2025^a	0.9264	3.0950^a	0.9002	2.9875^a	0.9350	1.4291^b	0.6377	48.65*	0.000
High context	2.4165^a	0.6743	2.3666^a	0.6877	2.3565^a	0.6770	2.2211^a	0.7271	–	0.456
Low context	3.0565^a	0.8045	3.0234^a	0.7924	2.9834^b	0.8187	2.0121^b	0.5724	22.92*	0.000
Composite score	3.0839^a	1.2473	2.9800^b	1.2245	2.9517^b	1.2331	2.1536^c	1.0394	159.28*	0.000

Notes: Means with different superscripts are statistically different from one another;* $p < 0.05$.

prominently in German than Latin American websites, which was contrary to the hypothesized direction.

Power distance.   German websites did not depict lower levels of power distance than US websites (H5a not supported). Approximately equal cues and numbers of attributes regarding company hierarchy information, pictures of CEOs, proper titles and vision statement were identified on German websites and UK sites, thus H5b was partially supported. As for H5c German sites portrayed higher levels of PD than Latin America, which again was contrary to expectations.

High and low context.   In terms of Hall's (1976) contextual dimensions, H6a was strongly supported, with German, US and UK websites forming a cluster of equivalent high low-context values compared to a statistically different and small low-context Latin American occurrence. Surprisingly, no major differences were found along the high context dimension, thus H6b was not supported.

Discussion and limitations

Within their home market, German companies' online websites adhered to cultural requirements. However, while on a general level their US and UK sites are in line with Hofstede's and Hall's cultural value frameworks, detailed cultural adaptation was missing. Thus our findings suggest a more standardized approach to an online web presence in these markets. While this approach might serve German companies well in the US and the UK markets, we observed a standardization approach that was rather crude in the Latin American markets, and gave the impression of 'cultural alienation'. Although this may be because the Latin American markets are of lesser economic importance, it might nevertheless place German companies in a weak competitive position against Latin American or US competitors. Quite clearly, German companies have not yet reached the minds of customers in the ever-increasing, economically important Latin America market (Singh and Baack, 2004).

To resolve this issue, more communication and flow of technological knowledge from German parent organizations and international counterparts in terms of developing and operating dynamic, culturally appropriate websites is suggested. An integrative network/transnational model can help in developing specialized resources and capabilities among the subsidiaries (for example, cultural acquaintance) while maintaining a close relationship with the parent organization.

A limitation of this chapter is that we did not control for different product or industry segments. For future work, it will be interesting to investigate whether there are different approaches to website standardization or adaptation, depending on b-to-b, b-to-c or other contextual factors. We also found that larger German companies derived greater web efficiencies from a somewhat more standardized approach to website management. It would be worthwhile to compare these size effects further in future studies. There is also a notion that the cultural framework used as an operationalization cannot only be helpful in cultural value depiction but also in the general approach to website design. This should be explored in the context of web-quality analyses. Finally, it is important to note that the Internet is a fast-moving and constantly evolving medium. A more dynamic approach to website analysis is encouraged, probably involving shorter time-periods regarding the analysis and frequent revisits to websites.

References

Agrawal, Madhu (1995) 'Review of a 40-Year Debate in International Advertising: Practitioner and Academician Perspectives to the Standardization/Adaptation Issue', *International Marketing Review*, 12(1), 26–48.

Alashban, Aref A., Linda A. Hayes, George M. Zinkhan and Anne L. Balazs (2002) 'International Brand-Name Standardization/Adaptation: Antecedents and Consequences', *Journal of International Marketing*, 10(3), 22–48.

Alvarez, Maria Gabriela, Leonard R. Kasday and Steven Todd (1998) *How We Made the Web Site International and Accessible: A Case Study* (Holmdel, NJ: AT&T Labs).

Bernard, Michael (2003) 'Criteria for Optimal Web Design (Designing for Usability)', http://psychology.wichita.edu/optimalweb/print.htm (Accessed 21 March 2006).

Berthon, Pierre, Leyland Pitt and Richard T. Watson (1996) 'Marketing Communication and the World Wide Web', *Business Horizons*, 39(5), 24–32.

Buzzell, Robert D. (1968) 'Can You Standardize Multinational Marketing?', *Harvard Business Review*, 46(6), 102–13.

Carr, Nicholas G. (2003) 'It Doesn't Matter', *Harvard Business Review*, 81(5), 41–9.

Cavusgil, S. Tamer and Shaoming Zou (1994) 'Marketing Strategy–Performance Relationships: An Investigation of the Empirical Link in Export Market Ventures', *Journal of Marketing*, 58(1), 1–21.

Cavusgil, S. Tamer, Shaoming Zou and G. M. Naidu (1993) 'Product and Promotion Adaptation in Export Ventures: An Empirical Investigation', *Journal of International Business Studies*, 24(3), 479–506.

Chakraborty, Goutam, Prashant Srivastava and David L. Warren (2005) 'Understanding Corporate B2B Web Sites' Effectiveness from North American and European Perspectives', *Industrial Marketing Management*, 34(5), 420–9.

Cho, Bongjin, Up Kwon, James W. Gentry, Sunkyu Jun and Fredric Kropp (1999) 'Cultural Values Reflected in Theme and Execution: A Comparative Study of US and Korean Television', *Journal of Advertising*, 28(4), 59–73.

Cyr, Dianne and Haizley Trevor-Smith (2004) 'Localization of Web Design: An Empirical Comparison of German, Japanese, and United States Web Site Characteristics', *Journal of the American Society for Information Science and Technology*, 55(13), 1199–208.

Douglas, Susan P. and Yoram Wind (1987) 'The Myth of Globalization', *Columbia Journal of World Business*, 22(4), 19–29.

Doz, Yves L. and C. K. Prahalad (1980) 'How MNCs Cope with Host Government Intervention', *Harvard Business Review*, 58(2), 149–57.

Dutta, Soumitra and Amit Jain (2004) 'The Networked Readiness Index: 2003–2004 Overview and Analysis Framework' World Development Report, www.weforum.org/pdf/Gcr/GITR_2003_2004/Framework_Chapter.pdf (Accessed 21 March 2005).

Evans, Philip B. and Thomas S. Wurster (1997) 'Strategy and the New Economics of Information', *Harvard Business Review*, 75(5), 70–82.

Fatt, Arthur C. (1967) 'The Danger of "Local" International Advertising', *Journal of Marketing*, 31(1), 60–2.

Federal Statistical Office Germany (2004) 'Ranking of Germany's Trading Partners in Foreign Trade', www.destatis.de/cgi-bin/ausland_suche_e.pl (Accessed 21 March 2006).

Fink, Dieter and Ricky Laupase (2000) 'Perceptions of Web Site Design Characteristics: A Malaysian/Australian Comparison', *Internet Research*, 10(1), 44–55.

Green, Robert T., William H. Cunningham and Isabella C. M. Cunningham (1975) 'The Effectiveness of Standardized Global Advertising', *Journal of Advertising*, 4(3), 25–30.

Gudykunst, William B. (1998) *Bridging Differences: Effective Intergroup Communication* (Newbury Park, Calif.: Sage).

Hall, Edward Twitchell (1976) *Beyond Culture* (New York: Anchor Press).

Han, Sang-Pil and Sharon Shavitt (1994) 'Persuasion and Culture: Advertising Appeals in Individualistic and Collectivistic Societies', *Journal of Experimental Social Psychology*, 30(4), 326–50.

Hodges, Mark (1997) 'Preventing Culture Clash on the World Wide Web', *MIT's Technology Review*, 100(8), 18–19.

Hofstede, Geert (1984) *Culture's Consequences: International Differences in Work-Related Values*, Cross-Cultural Research and Methodology Series (Newbury Park, Calif.: Sage).

Hofstede, Geert (1991) *Cultures and Organisations: Software of the Mind* (Maidenhead: McGraw-Hill).

Internet World Stats (2005) 'Internet Usage Statistics – the Big Picture', Miniwatts Marketing Group, http://www.internetworldstats.com (Accessed 21 March 2006).

Jain, Subhash C. (1989) 'Standardization of International Marketing Strategy: Some Research Hypotheses', *Journal of Marketing*, 53(1), 70–9.

Johansson, Johny K. (1994) 'The Sense of "Nonsense": Japanese TV Advertising', *Journal of Advertising*, 23(1), 18–25.

Kambil, Ajit (1995) 'Electronic Commerce: Implications of the Internet for Business Practice and Strategy', *Business Economics*, 30(4), 27.

Katsikeas, Constantine S., Saeed Samiee and Marios Theodosiou (2006) 'Strategy Fit and Performance Consequences of International Marketing Standardization', *Strategic Management Journal*, 27(9), 867–90.

Kotha, Suresh (1998) 'Competing on the Internet: The Case of Amazon.com', *European Management Journal*, 16(2), 212–22.

Kustin, Richard Alan (2004) 'Marketing Mix Standardization: A Cross Cultural Study of Four Countries', *International Business Review*, 13(5), 637–49.

Laughlin, Jay L., D. Wayne Norvell and David M. Andrus (1994) 'Marketing Presbyopia', *Journal of Marketing Theory and Practice*, 2(4), 1–10.

Levitt, Theodore (1960) 'Marketing Myopia', *Harvard Business Review*, 38(4), 45–56.

Levitt, Theodore (1983) 'The Globalization of Markets', *Harvard Business Review*, 61(3), 92–102.

Lim, Kai H., Kwok Leung, Choon Ling Sia and Matthew K. O. Lee (2004) 'Is eCommerce Boundary-less? Effects of Individualism–Collectivism and Uncertainty Avoidance on Internet Shopping', *Journal of International Business Studies*, 35(6), 545–59.

Lopez-Claros, Augusto and Soumitra Dutta (2005) *The Global Information Technology Report 2005–2006* [Online]. World Economic Forum. Available: http://www.weforum.org/gitr [2006, 21 March].

Luna, David, Laura A. Peracchio and Maria D. de Juan (2002) 'Cross-cultural and Cognitive Aspects of Web Site Navigation', *Journal of the Academy of Marketing Science*, 30(4), 397–410.

Lynch, Patrick D. and John C. Beck (2001) 'Profiles of Internet Buyers in 20 Countries: Evidence for Region-Specific Strategies', *Journal of International Business Studies*, 32(4), 725–48.

Lynch, Patrick D., Robert J. Kent and Srini S. Srinivasan (2001) 'The Global Internet Shopper: Evidence from Shopping Tasks in Twelve Countries', *Journal of Advertising Research*, 41(3), 15–23.

Marcus, Aaron and Emilie West Gould (2000) 'Crosscurrents – Cultural Dimensions and Global User-Interface Design', *Interactions*, 7(4), 32–46.

Mueller, Barbara (1992) 'Standardization vs. Specialisation: An Examination of Westernization in Japanese Advertising', *Journal of Advertising Research*, 32(1), 15–24.

Okazaki, Shintaro (2004) 'Do Multinationals Standardise or Localise? The Cross-cultural Dimensionality of Product-based Web Sites', *Internet Research: Electronic Networking Applications and Policy*, 14(1), 81–94.

Okazaki, Shintaro and Javier Alonso Rivas (2002) 'A Content Analysis of Multinationals' Web Communication Strategies: Cross-cultural Research Framework and Pre-Testing', *Internet Research: Electronic Networking Applications and Policy*, 12(5), 380–90.

Onkvisit, Sak and John J. Shaw (1987) 'Standardized International Advertising: A Review and Critical Evaluation of the Theoretical and Empirical Evidence', *Columbia Journal of World Business*, 22(3), 43–54.

Papavassiliou, Nikolaos and Vlasis Stathakopoulos (1997) 'Standardization versus Adaptation of International Advertising Strategies: Towards a Framework', *European Journal of Marketing*, 31(7), 504–27.

Petersen, Bent, Lawrence S. Welch and Peter W. Liesch (2002) 'The Internet and Foreign Market Expansion by Firms', *Management International Review*, 42(2), 207–21.

Quelch, John A. and Edward J. Hoff (1986) 'Customizing Global Marketing', *Harvard Business Review*, 64(3), 59–64.

Quelch, John A. and Lisa R. Klein (1996) 'The Internet and International Marketing', *Sloan Management Review*, 37(3), 60–75.

Ricks, David A. (1999) *Blunders in International Business*, 3rd edn (Oxford: Blackwell).

Riquelme, Hernan (2002) 'Commercial Internet Adoption in China: Comparing the Experience of Small, Medium and Large Businesses', *Internet Research*, 12(3), 276–86.

Rutigliano, Anthony J. (1986) 'The Debate Goes On: Global vs. Local Advertising', *Management Review*, 75(6), 27–31.

Samiee, Saeed (1998) 'The Internet and International Marketing: Is There a Fit?', *Journal of Interactive Marketing*, 12(4), 5–21.

Singh, Nitish (2002) 'Analyzing Cultural Sensitivity of Websites: A Normative Framework', *Journal of Practical Global Business*, April, 32–53.

Singh, Nitish and Daniel W. Baack (2004) 'Web Site Adaptation: A Cross-Cultural Comparison of U.S. and Mexican Web Sites', *Journal of Computer-Mediated Communication*, 9(4). Available at http://www.ascusc.org/jcmc/vol9/issue4/singh_baack.html.

Singh, Nitish and Paul D. Boughton (2002) 'Measuring Web Site Globalization: A Cross-sectional Country and Industry Level Analysis', *American Marketing Association Conference Proceedings*, 13, 302–3.

Singh, Nitish, Olivier Furrer and Massimiliano Ostinelli (2004) 'To Localize or to Standardize on the Web: Empirical Evidence from Italy, India, Netherlands, Spain, and Switzerland', *Multinational Business Review*, 12(1), 69–87.

Singh, Nitish, Hongxin Zhao and Xiaorui Hu (2003) 'Cultural Adaptation on the Web: A Study of American Companies' Domestic and Chinese Websites', *Journal of Global Information Management*, 11(3), 63–80.

Singh, Nitish, Hongxin Zhao and Xiaorui Hu (2005a) 'Analyzing the Cultural Content of Web Sites: A Cross-national Comparision of China, India, Japan, and US', *International Marketing Review*, 22(2), 129–46.

Singh, Nitish, Vikas Kumar and Daniel Baack (2005b) 'Adaptation of Cultural Content: Evidence from B2C E-Commerce Firms', *European Journal of Marketing*, 39(1/2), 71–86.

Sinkovics, Rudolf R. and Elfriede Penz (2005) 'Empowerment of SME Websites–Development of a Web-empowerment Scale and Preliminary Evidence', *Journal of International Entrepreneurship*, 3(4), 303–15.

Subramaniam, Mohan and Kelly Hewett (2004) 'Balancing Standardization and Adaptation for Product Performance in International Markets: Testing the Influence of Headquarters–Subsidiary Contact and Cooperation', *Management International Review*, 44(2), 171–94.

Theodosiou, Marios and Leonidas C. Leonidou (2003) 'Standardization versus Adaptation of International Marketing Strategy: An Integrative Assessment of the Empirical Research', *International Business Review*, 12(2), 141–71.

Triandis, Harry C. (1994) *Culture and Social Behaviour* (New York: McGraw-Hill).

Tsikriktsis, Nikos (2002) 'Does Culture Influence Web Site Quality Expectations? An Empirical Study', *Journal of Service Research*, 5(2), 101–12.

Walters, Peter G. P. (1986) 'International Marketing Policy: A Discussion of the Standardization Construct and Its Relevance for Corporate Policy', *Journal of International Business Studies*, 17(2), 55–69.

Walters, Peter G. P. and Brian Toyne (1989) 'Product Modification and Standardisation in International Markets: Strategic Options and Facilitating Policies', *Columbia Journal of World Business*, 24(4), 37–44.

World Economic Forum (2005) 'Networked Readiness Index', Wikipedia, http://en.wikipedia.org/wiki/Networked_readiness_index (Accessed 4 August 2006).

Yamin, Mohammad and Rudolf R. Sinkovics (2006) 'Online Internationalisation, Psychic Distance Reduction and the Virtuality Trap', *International Business Review*, 15(4), 339–60.

Yip, George and Anna Dempster (2005) 'Using the Internet to Enhance Global Strategy', *European Management Journal*, 23(1), 1–13.

Yip, George S. (1989) 'Global Strategy . . . In a World of Nations?', *Sloan Management Review*, 31(1), 29–41.

Yip, George S. (2000) 'Global Strategy in the Internet Era', *Business Strategy Review*, 11(4), 1–13.

Zahir, Sajjad, Brian Dobing and M. Gordon Hunter (2002) 'Cross-cultural Dimensions of Internet Portals', *Internet Research*, 12(3), 210–20.

Zandpour, Fred, Veronica Campos, Jeolle Catalano, Cypress Chang, Young Dae Cho, Renee Hoobyar, Shu-Fang Jiang, Man-Chi Lin, Stan Madrid, Holly Scheideler and Susan Titus Osborn (1994) 'Global Reach and Local Touch: Achieving Cultural Fitness in TV Advertising', *Journal of Advertising Research*, 34(5), 35–63.

Zou, Shaoming and S. Tamer Cavusgil (1996) 'Global Strategy: A Review and an Integrated Conceptual Framework', *European Journal of Marketing*, 30(1), 52–69.

Zou, Shaoming, David M. Andrus and D. Wayne Norvell (1997) 'Standardization of International Marketing Strategy by Firms from a Developing Country', *International Marketing Review*, 14(2), 107–23.

13
The Geographical Dimension: A Missing Link in the Internationalization of 'Born Global' Firms?

Olli Kuivalainen, Sanna Sundqvist and Per Servais

Introduction

Being an international firm is often seen as a critical ingredient of the firm's strategy for achieving growth, sustainable competitive advantage and superior performance. In many earlier studies, increased multinationality is seen as a good thing for a firm's performance, and this can be seen as a basic premise for international business (see, for example, Contractor *et al.*, 2003; Tallman and Li, 1996). It has even been argued that the success or failure of a firm, in the current competitive and uncertain environment, will depend on whether it can compete effectively on a global scale (Ohmae, 1989). Consequently, there is an increasing amount of evidence of firms that, despite being small or at an early stage in their development and possessing limited resources, are aiming towards rapid internationalization and are anxious to participate in the global competitive landscape. Such firms, often called 'born globals' (see, Rennie, 1993), are influenced by the globalization of markets and customer needs, and the impact of new communication and transportation technologies (Knight and Cavusgil, 1996, 2004). However, while internationalization and its effect on firm performance have been the subject of intensive research since the mid-1970s (see, for example, Grant, 1987; Li, 2005; Ruigrok and Wagner, 2003; Sullivan, 1994) there is a scarcity of empirical research to determine whether rapid, accelerated internationalization is in fact profitable. To date, we know surprisingly little regarding the performance consequences of internationalization in the case of small and medium-sized firms. There are several reasons for this. First, detailed information regarding small firms is hard to obtain, as many firms do not publish all the records focusing on the infant years of their existence and internationalization (see Lu and Beamish 2001), and the correspondingly numerous risks associated with international business (see

Cavusgil, 2006) lead to the fact that a large number of firms that embarked on internationalization during their early years fail, and cannot be researched without difficulty. Second, the extant born global research has been mainly descriptive in nature; the aim of most of the published studies has been to understand and interpret the principal reasons that have led to the emergence of early and rapidly internationalizing firms (see the review in Rialp *et al.*, 2005). Third, the few existing results are contradictory or ambiguous – for example the results of a survey conducted in the USA did not prove that rapid internationalization led to better performance, but it did not prove the reverse either (Bloodgood *et al.*, 1996).

In this chapter, another reason is observed. Despite the recent increase in born global studies, there has been little research into the effect on performance of operating across many countries, regions and continents. In many born global studies it is only noted that born globals often derive their turnover from multiple countries (see, for example, Knight and Cavusgil, 2004; Oviatt and McDougall, 1994) but this 'country effect' has been excluded from the empirical work. In many cases, the degree of the internationalization is only studied by measuring the share of the firm's sales generated from foreign markets (see the review by Zahra and George, 2002). While there are some exceptions (see, for example, Reuber and Fischer 1997; Zahra *et al.*, 2000) it seems that multinationality or global diversity of firms' operations, which can be seen as core issues in international management research, are missing from the born global studies.

Thus, this chapter first explores the concept of global diversity in the case of born global firms; second tries to identify the drivers of global diversification strategy; and finally looks at the performance consequences. In the following pages we first summarize the extant literature on the degree and scope of rapid internationalization. Then we present the results from an empirical study conducted in Finland, and finally conclude with the discussion and implications for future research, as well as suggestions for practising managers.

Literature review: degree and scope of rapid internationalization

The focus of earlier research on born globals can be divided into four main dimensions: (i) extent/degree of internationalization of a firm's sales; (ii) speed/pace between the year of foundation and first foreign sales; (iii) scope (that is, number and diversity of target countries/markets; and (iv) drivers or antecedents (for example, managerial and firm characteristics) of rapid internationalization (see, for example, Bloodgood *et al.*, 1996; Knight and Cavusgil, 1996; Zahra and George, 2002). In most of the studies focusing on born globals, the firm's degree of internationalization is measured by studying the amount of turnover derived from international operations (see dimension (i) above). In this, the criterion has often been 25 per cent of the total turnover (see, for example, Knight and Cavusgil, 1996; Madsen *et al.*, 2000; Moen, 2002). It seems that this criterion[1] is widely accepted in the field.

In the case of scope, as mentioned above, there are no exact definitions. In their seminal study, Oviatt and McDougall (1994) present a typology of small, rapidly internationalizing firms based on the number of countries involved and the amount of co-ordination of value chain activities. Born global firms (or international new ventures) are defined 'as business organisations that from inception, seek to derive significant competitive advantage from the use of resources and the sale of outputs in multiple countries' (Oviatt and McDougall, 1994, p. 49). However, in this excellent paper Oviatt and McDougall do not present detailed criteria for empirical studies regarding market scope, and the same can be said of the above-mentioned paper by Knight and Cavusgil (1996).

Consequently, while the born global concept states implicitly that a firm following that strategy or pathway must begin to operate in multiple countries on international markets almost from its inception (see, for example, Oviatt and McDougall, 1994), we do not know much about the role of market scope, whether it be regional, continental or global. Few of the studies that have examined the role of countries or regions in the internationalization process of small entrepreneurial firms include, for example, Zahra *et al.* (2000) and Reuber and Fischer (1997). Zahra *et al.* (2000) studied US-based new ventures that diversified internationally (measured by five indicators including, for example, number of export countries and geographical diversity). They found that some dimensions of diversity have positive effects on performance, and some have negative results or none at all. However, there is still only limited information available. First, both of the above-mentioned studies (as are many others) are based on samples of North American high-technology firms, and second, some measures used are context specific and do not allow easy comparisons. For example, Reuber and Fischer (1997) focus on Canadian software firms and look at the degree of internationalization (DOI) partly by dividing export regions into three – namely, Canada, North America and 'outside North America'.

It seems that this important area of international business research has mainly been a topic when the centre of attention is on large multinational corporations (see, for example, Li, 2005; Rugman and Verbeke, 2004). What is implicitly clear, however, is that a born global pathway should contradict the behavioural incremental internationalization models (see, for example, Bilkey and Tesar, 1977; Johanson and Vahlne, 1977, 1990). Thus a market selection and the beginning of the international operations of a born global firm should not only begin from neighbouring countries and/or with the markets with low psychic distance.

A seminal study conducted by Ayal and Zif (1979) hypothesized that for each firm there may be an optimum number of export countries; the strategies for reaching this optimum may be different and even oppose each other. It is evident that, for some firms, a suitable strategy is to seek as many markets as possible (see Piercy, 1981). For example, according to Hitt *et al.* (1994) a broad international market scope increases the chances of survival. However, for some firms the key to successful operations is to focus on a few main markets (see Tessler, 1980). Even for a large firm, a wide market scope may not always be the best solution; the extent to which firms are able to realize the benefits of rapid internationalization is constrained by their capacity to handle and absorb the complexities that accompany international expansion (Vermeulen and Barkema, 2002). There are differences; for example, for firms which can, and actively seek to, operate internationally online, the sequencing of foreign market entry is likely to be much more 'time-compressed' compared with traditional internationalization (Yamin and Sinkovics, 2006).

Born-global pathways: research questions

There are also different strategies for becoming a born global or, to put it another way, born global strategies. Some young internationalizing firms diversify rapidly geographically and increase their market scope, while some only export intensively to a few carefully chosen countries. It seems that there are 'truly born globals', which operate in numerous distant markets/regions, and other 'born globals' that fulfil, for example, the common turnover criteria by generating more than 25 per cent of their sales from exporting but do not have a global geographical spread or scope.

Given a certain pace for a firm's internationalization process, a higher geographical spread to numerous markets may imply that a firm operating in a large number of regions, and thus following a true 'born global' pathway, has to learn more about unique target market settings and, if more successful, it may have to possess more born global qualities – that is, qualities normally associated with born global firms. These internal drivers of rapid internationalization may include, for example, international entrepreneurial orientation. According to Lumpkin and Dess (2001, p. 429) the term 'entrepreneurial orientation' refers to 'the strategy-making processes and styles of firms that engage in entrepreneurial activities'. The associated dimensions of this construct are autonomy, innovativeness, risk-taking, proactiveness and competitive aggressiveness (see, for example, Lumpkin and Dess, 2001) and they can be seen as important drivers of born global strategy. For example, rapid internationalization is often based on entrepreneurial risk-taking and proactive behaviour.

Rapid internationalization and simultaneous global diversification may also be responses to the environmental situation. Strategic behaviour theory suggests that a firm conducts business by whichever mode maximizes profits through the improvement in the competitive situation (see, for example, Porter, 1991). This theory can also be seen to explain the internationalization process (see

Knight, 2001) and increase in the market scope. Our propositions are that there are differences between born global firms that have a wider scope (global diversity) and born global firms operating only in the regions close to their home markets in terms of:

- internal drivers of internationalization such as international entrepreneurial orientation;
- external drivers that may make a firm internationalize (for example environmental turbulence); and
- export performance.

Our research questions are partly explorative by nature. In the following empirical part of the study we present empirical results from Finland in order to develop our understanding of the phenomenon of born globals.

Methodology

Sample and data collection

To explore the role of global diversity of born global firms, a postal survey of Finnish exporters was conducted, using measures culled from the literature. The sample frame was Kompass Finland's (SKOD, 1998) listing of Finnish exporters with fifty or more employees (a total of 1,205 firms). The target respondent was the export marketing manager. Each firm was contacted by telephone to determine eligibility, and a questionnaire, an explanatory cover letter and a postage-paid reply envelope were sent to eligible firms. Ten days after the first mailing, a reminder card was sent to each non-respondent. Another questionnaire was mailed seven days later to all remaining non-respondents, together with an appropriate letter and a reply envelope. Of the 1,205 contacts in the sample, 237 were ineligible because the firm had never exported, no longer exported, or was listed more than once. Of the 968 questionnaires mailed, 783 usable responses were obtained, making a response rate of 81 per cent. Non-response bias was not considered to be an issue for two reasons (i) the relatively high response rate reduces the risk of response bias; and (ii) a comparison between early and late respondents on all variables of interest demonstrated that no significant differences existed (Armstrong and Overton, 1977).

Measures

Born globalness

Born global firms were defined as firms whose share of international sales was at least 25 per cent, and who had internationalized within three years of establishment, as suggested by, for example, Knight and Cavusgil (1996) and Knight *et al.* (2004).

Scope of internationalization

Respondents were asked to evaluate the dispersion of their export sales by dividing their export sales (percentages) into the following eight regions: Australia and Oceania, Asia, Russia and Baltic countries, Western Europe, Northern America, South America, Africa and the Middle East, and Eastern Europe.

Entrepreneurial orientation

Three dimensions of entrepreneurial orientation were studied: proactiveness, risk-taking and competitive aggressiveness. The proactiveness was captured using Jambulingam *et al.*'s (2005) proactiveness scale. The scale was adapted for an international business (exporting) context and reflected the extent to which managers seized opportunities in anticipation of future market conditions. We captured the degree to which managers take risks, using items drawn from Jambulingam *et al.*'s (2005) risk-taking scale. The adapted scale gauges the role of risk-taking as a part of a firm's internationalization strategy. Competitive aggressiveness measure was based on items from Narver and Slater's (1990) competitor orientation scale, and Jaworski and Kohli's (1993) market responsiveness scale.

Environmental turbulence

Three environment measures were adapted from Cadogan *et al.* (1998), who used measures developed initially by Jaworski and Kohli (1993), and subsequently modified them for use in an export setting. Specifically, the scales captured technological turbulence (to capture changes and opportunities occurring in the firm's export markets as a result of technology), competitor intensity (which assessed the ease with which a firm can differentiate in its export markets), and customer dynamism (capturing changes in export customer preferences and needs, customer demand and market growth).

Export performance

Following the recommendations of Cavusgil and Zou (1993) and Matthyssens and Pauwels (1996), among others, we measured aspects of the firm's export sales and export profits. Our 'sales performance' measure contained items to capture (i) the firm's sales growth relative to the industry average, (ii) the firm's degree of satisfaction with its export volume, (iii) the firm's degree of satisfaction with its market share in its export markets; and (iv) the firm's degree of satisfaction with its rate of new market entry. Our 'profit performance' measure captured (i) the firm's degree of satisfaction with its export profits over the previous three years; and (ii) an overall assessment of the profitability of the firm's exporting operations during the previous financial year. The third performance measure 'efficiency performance' captured (i) the ratio of the firm's total annual export sales turnover to the total number of employees working in the firm; and (ii) the ratio of the firm's total annual export sales turnover to the total number of countries to which the firm exports.

Control variables

As the number of export countries is often included in earlier studies as a proxy for the scope of internationalization, we added export intensity as a control variable.

Analyses and results

Our data consisted of 783 exporting firms, of which 185 were chosen for analyses as they fulfilled the above-mentioned Knight and Cavusgil (1996) and Knight *et al.* (2004) criteria for born global firms. Firms had been in business for 38 years on average, and had operated in international markets for 37 years. Born global firms derived almost 80 per cent of their turnover from international markets. The average size of the firm was 1,586 employees and average turnover €378 million (see Table 13.1).

It was proposed that even born global companies differ in their global diversity. Thus it was first explored whether different global diversity profiles could be identified. We used cluster analysis, following the guidelines of Hair *et al.* (1998), to search for these global diversity profiles. Hierarchical cluster analysis was performed using the eight measures describing the scope of internationalization as the clustering variates. The similarity between cases was measured by squared Euclidean distance, and the clusters were formed using the Ward method. Based on the increase in the agglomeration coefficient, the appropriate number of clusters was six. The cluster solution was validated by using the quick clustering procedure. The solution derived from quick clustering did not differ from the results of hierarchical clustering. The global diversity profiles are shown in Table 13.2, and described in Table 13.3.

Table 13.1 Basic information on the respondents

	Mean	Std. Dev.	Min.	Max.
Turnover (millions €)	378.52	1,588.69	2.02	13,455.03
Experience (years in business)	37.55	42.99	3	313
Full-time employees	1,586.37	6,088.50	12	44,500
International experience (years exporting)	37.19	43.08	3	313
Share of foreign turnover (%)	74.70	22.08	25	100
Number of countries exported to	33.45	30.73	1	200

Table 13.2   Global diversity clusters of respondents

Percentage of:	Geographical clusters					
	1	2	3	4	5	6
Exports to Australia	0.00	0.00	1.06	2.48	3.65	8.52
Exports to Asia	0.00	4.50	4.33	8.04	8.37	23.22
Exports to Russia and Baltic countries	94.29	5.50	2.16	12.09	3.22	7.91
Exports to Western Europe	1.43	26.25	83.94	58.71	39.06	24.13
Exports to Northern America	0.00	0.50	3.26	5.22	34.30	11.65
Exports to South America	0.00	3.25	1.84	1.36	3.74	7.22
Exports to Africa and Middle East	0.00	0.75	1.53	5.17	2.63	8.17
Exports to Eastern Europe	4.29	59.25	1.91	6.31	4.67	8.00

Table 13.3   Description of global diversity clusters

Cluster	Description
Cluster 1 (N = 7)	Cluster 1 consists of firms that are very dependent on their exports to Russia and Baltic countries
Cluster 2 (N = 4)	In Cluster 2 there are firms that export mainly to Europe and home regions
Cluster 3 (N = 60)	Firms in this cluster export mainly to Western Europe and only occasionally to other countries/regions
Cluster 4 (N = 42)	Firms in Cluster 4 export mainly to Western Europe, but also some to Russia and Baltic countries as well as to Asia
Cluster 5 (N = 27)	Firms in Cluster 5 export to the most industrialized regions of the world – that is, to Western Europe and North America
Cluster 6 (N = 23)	Firms in Cluster 6 can be described as exporters whose target markets are distributed equally all over the world.

As proposed, born global firms seem to differ in their global diversity. Consequently, the drivers of these global diversities should be identified. Therefore, as discussed earlier, the effects of internal factors such as risk-taking, proactiveness and competitive aggressiveness, were explored. In order to test for unidimensionality all of the entrepreneurial orientation items were subjected to an exploratory factor analysis (principal factor analysis with varimax rotation). Items that cross-loaded at higher than the 0.40 level, or which did not load on any factor above the 0.40 level were eliminated from the scale. The final score for each entrepreneurial orientation scale was an average of all the items included in the scale. Appendix Table 13.A1 on page 237 shows the reliability statistics of the multi-item scales used in regression analyses. The reliability assessments show that all the scales exceeded the recommended level of 0.70 (Nunnally, 1978). (See Appendix Table 13.A2 on page 237 for principal axis factor analysis of proactiveness, risk-taking and competitive aggressiveness scale items.)

Table 13.4 Internal drivers of global diversity

Independent variable	B	Beta	T	Sig.
Risk-taking	−0.028	−0.041	−0.537	0.592
Proactiveness	0.145	0.072	2.016	0.045**
Competitive aggressiveness	−0.267	−0.208	−2.458	0.015**
Control variables				
Export Intensity (number of countries)	0.012	0.282	3.714	0.000***

Notes: Dependent variable–global diversity, R^2=0.11; *sig. < 0.01; **sig. < 0.05; ***sig. < 0.05.

In order to explore the drivers of global diversity, linear regression analysis was applied. On the basis of the analysis (see Table 13.4), it seems that only proactiveness is a significant driver of global diversity within our sample. Competitive aggressiveness appeared to have a reverse effect on global diversity (sig. 0.015), indicating that firms with more focused competitive strategies aiming at outperforming competitors apply a more concentrated international strategy. Surprisingly, risk-taking did not contribute towards more global diversity strategy in international business. In the next stage of the analysis, factors external to the firm were also explored. Multi-item scales (see the development of entrepreneurial orientation measures as a description of the development procedure) were applied to measure the three dimensions of the firm's export environment (technological turbulence, competitor intensity and customer dynamism). In order to study the effect of environmental turbulence on global diversity, linear regression analysis was applied.

Based on the regression results, it seems that external drivers also have an effect on global diversity. The turbulence of the technological environment appears to have the strongest effect on global diversity. Results show that firms operating in technologically turbulent environments appear to favour a more global export strategy. In contrast, firms operating in environments where customer needs are changing rapidly, or in environments where the competitive environment is hostile, favour operating in more local and closed export markets (see Table 13.5).

We were also interested to see whether a truly born global strategy (that is, high global diversity) leads to a better export performance. Thus we explored whether there exist differences across different global diversity strategies in

Table 13.5 External drivers of global diversity

Independent variable	B	Beta	T	Sig.
Competitor environment turbulence	−0.191	−0.166	−2.132	0.035**
Customer environment turbulence	−0.157	0.086	−1.836	0.068*
Technology environment turbulence	0.188	0.223	2.836	0.005***

Notes: Dependent variable–global diversity; R^2 = 0.09; *sig. < 0.01; **sig. < 0.05; ***sig. < 0.005.

Table 13.6 Performance differences across firms with different global diversity profiles

	Cluster 1			Cluster 2			Cluster 3		
	N	Mean	Std. Dev.	N	Mean	Std. Dev.	N	Mean	Std. Dev.
Sales performance	7	5.64	2.47	4	6.94	1.71	57	6.97	1.35
Profit performance	7	6.07	2.30	4	6.38	1.38	59	7.25	1.39
Efficiency performance	7	0.77	0.72	4	0.22	0.54	55	0.05	0.36

	Cluster 4			Cluster 5			Cluster 6		
	N	Mean	Std. Dev.	N	Mean	Std. Dev.	N	Mean	Std. Dev.
Sales performance	38	6.78	1.51	27	7.06	1.20	21	6.26	1.53
Profit performance	41	6.88	1.63	27	7.41	1.77	23	6.89	1.99
Efficiency performance	42	0.20	1.11	27	0.08	0.66	23	0.05	0.37

	ANOVA	
	F	Sig.
Sales performance	1.794	0.118
Profit performance	1.169	0.327
Efficiency performance	1.482	0.199

export performances. Analysis of variance was applied to reveal the differences. However, contrary to our expectations, no significant differences in export performances across different types of born global firms were identified (see Table 13.6). Table 13.6 reveals that firms which export to the most industrialized regions of the world, that is to Western Europe and North America (Cluster 5) seem to have the highest sales and profit performance, whereas firms operating only in Russian and the Baltic markets report relatively lower export performance indicators. However, as the analysis of variance results show, these differences are not statistically significant.

Discussion

The earlier research has shown that there is enormous variety in the internationalization paths of small firms (see, for example, Jones, 1999). Extant studies have also shown that born global firms tend to differ from their more slowly internationalizing counterparts in the degree of the various drivers of internationalization, such as entrepreneurial orientation, the motivation of the management for making the firm international, strategy and so on. (see, for example, Knight and Cavusgil, 1996, 2004).

In our study we tried to find differences among born globals – that is, between 'truly born globals' operating in diverse markets around the globe and other born global firms that could be called 'born internationals' and which make their foreign sales in regions close to the home market. As the results from small open economies suggest, the criteria often associated with the work of Knight and Cavusgil may make most of the manufacturing small firms born globals (see Moen, 2002), and it was assumed that some variety should exist among the firms operating in different market clusters. However, we have to admit that our results partially contradict our expectations because, of the internal drivers tested only proactiveness had a statistically significant positive effect on a firm's global diversity, and competitive aggressiveness contributes negatively to the global diversity of the firm. When it comes to external drivers, environmental turbulence played a role in global diversity among rapidly internationalized firms. The firms that thought there were changes in technologies had spread their operations across a large number of regions. This may have occurred out of necessity; the market had to be penetrated in a short period of time because of a short product life-cycle, for example. Furthermore, the firms that considered there is customer turbulence were operating closer to the home market. This may be based on the fact that a shorter physical and psychic distance enables better interaction with one's customers, but we can only speculate on this. In general, it seems that the Finnish exporting firms that participated in the survey were adapting their operations to the environment, following their customers and focusing on market development.

Contrary to our expectations, significant differences in performance indicators among different types of born global firms could not be found. Established theory suggests a direct link between international expansion and performance (see, for example, Hitt *et al.* 1994, 1997), as it is believed that a broad international market scope stabilizes the firm's earnings (Caves, 1982), and increases the chances of survival (Hitt *et al.*, 1994). However, there are a few possible explanations for why this type of positive impact may not always occur in the case of rapidly internationalizing firms, one of them naturally being the amplification of uncertainty as a firm spreads to geographically and culturally diverse environments. In addition, 'superstitious learning' (see Levitt and March, 1988; Snyder and Cummings, 1998) may also lower performance in physically close markets. Yamin and Sinkovics (2006), focusing on the internationalization of online firms, point out that the so-called 'psychic distance paradox',[2] which is an example of superstitious learning, may also occur in physically close markets; in this case, managers may assume that devoting resources to learning about the chosen market is unnecessary, as the market is (mistakenly) seen to be too similar to the home market. It is tempting to think that this type of mistake may lead to diminishing performance consequences, even if a firm held the opinion that operating in the markets close to home in an age of anxiety was a better solution.

Furthermore, there are other issues that may have a significant role on the relationship between international market diversity and performance. For example, it has been found that product diversity moderates the relationship between international diversity and performance (Hitt *et al.*, 1997). Although this was of interest, because we did not explore inter-firm differences in product offering (or market/industry type, or customer type), we are hesitant to speculate about performance implications at this point.

A different explanation could also be that firms may also adapt their operations to the level of demand and expected returns. Having a certain financial stake-holder who expects an assured return for his/her investments may lead to a situation where firms become risk averse. In any case, the performance measures did not differ among the firms in our sample.

Accordingly, one of the main limitations in our study is the nature of the relationship between multinationality and performance. Some authors suggest that highly dispersed international operations increase management constraints and costs, and thus diminish performance benefits (see, for example, Grant, 1987). To overcome this measurement problem, longitudinal studies should be conducted. Furthermore, with longitudinal research design, it would be easier to study other benefits a firm can gain from foreign operations in an age of anxiety. These include, for example, learning benefits and knowledge-sharing with carefully chosen foreign partners (see Cavusgil, 2006).

Our sample can also be seen as being unrepresentative, as the firms studied in this chapter had all survived the stage of being a small firm. This may lead to 'survival bias' in the sample (see Vermeulen and Barkema, 2002). For example, as entrepreneurial orientation was not measured at the beginning of the export operations, interesting results might emerge if the firms that failed were compared to these experienced born global firms (which still exist). It is important to point out that future research should be designed to better facilitate such comparisons.

From the managerial perspective, it is important to notice the central role of environmental turbulence in the shaping of strategies related to the scope of international operations. The conducted environmental analysis and adjustment of the subsequent strategic plan to the various contextual situations may be a worthwhile idea.

Appendix 1

Table 13.A1 Scale reliability analyses

Scale	Cronbach's alpha	Number of items	Mean score	Std. Dev.
Internal drivers				
Proactiveness	0.856	3	6.056	1.509
Risk-taking	0.872	3	3.979	1.775
Competitive aggressiveness	0.750	4	4.930	1.095
External drivers				
Technological turbulence	0.896	4	4.334	1.504
Competitor intensity	0.748	6	4.946	1.024
Customer dynamism	0.772	5	4.306	1.174

Appendix 2

Table 13.A2 Principal axis factor analysis of proactiveness, risk-taking and competitive aggressiveness items

Item	Proactiveness	Risk-taking	Competitive aggressiveness	Communality
Export managers in this company usually take action in anticipation of future export market conditions	0.785			0.688
Export managers in this company try to shape our business environment to enhance our presence in the export market	0.794			0.686
Export managers in this company continually seek out new opportunities, because export market conditions are changing	0.740			0.642
Export managers in this company see taking a gamble as being part of its strategy for export success		0.809		0.709

Table 13.A2   (Continued)

Item	Proactiveness	Risk-taking	Competitive aggressiveness	Communality
Export managers in this company take above-average risks		0.839		0.707
Export managers in this company see taking chances as being an element of its export strategy		0.827		0.695
If a major competitor were to launch an intensive campaign targeted at their foreign customers, they would implement a response immediately			0.554	0.329
They are quick to respond to significant changes in their competitors' price structures in foreign markets			0.738	0.566
They take a long time to decide how to respond to export competitors' price changes*			0.486	0.273
They respond rapidly to competitive actions that threaten the company in its export markets			0.771	0.657
Eigenvalue	3.7	2.1	1.3	
Percentage variance	36.67	21.14	13.01	70.83**

Note: * Reverse-coded. ** Cumulative total.

Notes

1. One of the first commonly used criteria to define a born global firm is associated with Knight and Cavusgil (1996) and Knight (1997), and used, for example, in Knight *et al.* (2004): 'a born global is a firm that has reached a share of foreign sales of at least 25% after having started export activities within three years after its

birth'. In addition, they used the criterion that the firms had to be founded after the year 1977. In their more recent article, they notice that the 25 per cent cut-off ratio for exports is 'somewhat arbitrary' and 'established in light of the exploratory goals of the research' (Knight and Cavusgil, 2004, p. 133). However, the 25 per cent criterion is probably a figure that already means a firm needs to take its international operations seriously and internationalization is no longer sporadic. In this sense, this cut-off rate serves its purpose and was used when choosing the focal firms of this study.

2. According to O'Grady and Lane (1996, p. 309) the psychic distance paradox is that 'operations in psychically close countries are not necessarily easy to manage, because assumptions of similarity can prevent executives from learning about critical differences'. Another psychic distance paradox is that, if a firm perceives that psychic distance exists, it generally tends to adopt a more cautious approach and devote greater resources to learning about the market. This may lead to a situation where the subsequent performance implications may be positive, contrary to expectations (see Evans and Mavondo, 2002; Yamin and Sinkovics, 2006).

References

Armstrong, Scott J. and Terry S. Overton (1977) 'Estimating Non-Response Bias in Mail Surveys', *Journal of Marketing Research*, 14(3), 396–402.

Ayal, Igal and Jehiel Zif (1979) 'Marketing Expansion Strategies in Multinational Marketing', *Journal of Marketing*, 43(2), 84–94.

Barkema, Harry G. and Freek Vermeulen (1998) 'International Expansion through Start-Up or Acquisition: A Learning Perspective', *Academy of Management Journal*, 41(1), 7–26.

Bilkey, Warren J. and George Tesar (1977) 'The Export Behavior of Smaller Sized Wisconsin Manufacturing Firms', *Journal of International Business Studies*, 8(1), 93–8.

Bloodgood, James M., Harry J. Sapienza and James G. Almeida (1996) 'The Internationalization of New High-Potential U.S. Ventures: Antecedents and Outcomes', *Entrepreneurship: Theory and Practice*, 20(4), 61–76.

Cadogan, John W., Adamantios Diamantopoulos and Judy A. Siguaw (1998) 'Export Market-Oriented Behaviour, Their Antecedents, Consequences, and Moderating Factors: Evidence from the U.S. and the U.K.', in Per Andersson (ed.), *Marketing: Research and Practice; Vol. 2*, Proceedings of the Annual Conference of the European Marketing Academy, Stockholm, Sweden, 449–51.

Caves, Richard E. (1982) *Multinational Enterprise and Economic Analysis* (Cambridge University Press).

Cavusgil, S. Tamer (2006) 'International Business in the Age of Anxiety: Company Risks', Keynote presentation, 33rd AIB-UK Conference: Manchester, 6–8 April 2006. Available at http://www.mbs.ac.uk/aib2006.

Cavusgil, S. Tamer and Shaoming Zou (1993) 'Product and Promotion Adaptation in Export Ventures: An Empirical Investigation', *Journal of International Business Studies*, 24(3), 479–507.

Contractor, Farok J., Sumit K. Kundu and Chin-Chun Hsu (2003) 'A Three-Stage Theory of International Expansion: The Link between Multinationality and Performance in the Service Sector', *Journal of International Business Studies*, 34(1), 5–18.

Evans, Jody and Felix T. Mavondo (2002) 'Psychic Distance and Organizational Performance: An Empirical Examination of International Retailing Operations', *Journal of International Business Studies*, 33(3), 515–32.

Grant, Robert M. (1987) 'Multinationality and Performance among British Manufacturing Companies', *Journal of International Business Studies*, 22(3), 249–63.

Hair, Joseph F., Jr., Rolph E. Anderson, Ronald L. Tatham and William C. Black (1998) *Multivariate Data Analysis* (Englewood Cliffs, NJ Prentice-Hall).

Hitt, Michael A., Robert E. Hoskisson and Duane R. Ireland (1994) 'A Mid-Range Theory of the Interactive Effects of International and Product Diversification on Innovation and Performance', *Journal of Management*, 20(2), 297–326.

Hitt, Michael A., Robert E. Hoskisson and Hicheon Kim (1997) 'International Diversification: Effects on Innovation and Firm Performance in Product-Diversified Firms', *Academy of Management Journal*, 40(4), 767–98.

Jambulingam, Thanigavelan, Ravi Kathuria and William R. Doucette (2005) 'Entrepreneurial Orientation as a Basis for Classification within a Service Industry: The Case of Retail Pharmacy Industry', *Journal of Operations Management*, 23(1), 23–42.

Jaworski, Bernard J. and Ajay K. Kohli (1993) 'Market Orientation: Antecedents and Consequences', *Journal of Marketing*, 57(3), 53–70.

Johanson, Jan and Jan-Erik Vahlne (1977) 'The Internationalization Process of the Firm: a Model of Knowledge Development and Increasing Foreign Market Commitments', *Journal of International Business Studies*, 8(1), 23–32.

Johanson, Jan and Jan-Erik Vahlne (1990) 'The Mechanism of Internationalization', *International Marketing Review*, 7(4), 11–24.

Jones, Marian V. (1999) 'The Internationalization of Small High-Technology Firms', *Journal of International Marketing*, 7(4), 15–41.

Knight, Gary A. (1997) 'Emerging Paradigm for International Marketing: The Born-Global Firm', Doctoral dissertation, East Lansign, Mich.: Michigan State University.

Knight, Gary A. (2001) 'Entrepreneurship and Strategy in the International SME', *Journal of International Management*, 7(3), 155–71.

Knight, Gary A. and S. Tamer Cavusgil (1996) 'The Born Global Firm: A Challenge to Traditional Internationalization Theory', *Advances in International Marketing*, 8, 11–26.

Knight, Gary A. and S. Tamer Cavusgil (2004) 'Innovation, Organizational Capabilities, and the Born-Global Firm', *Journal of International Business Studies*, 35(2), 124–41.

Knight, Gary A., Tage K. Madsen and Per Servais (2004) 'An Inquiry into Born-Global Firms in Europe and the USA', *International Marketing Review*, 21(6), 645–65.

Levitt, Barbara and James G. March (1988) 'Organizational Learning', *Annual Review of Sociology*, 14, 319–40.

Li, Lei (2005) 'Is Regional Strategy More Effective than Global Strategy in the US Service Industries?', *Management International Review*, 45(1), 37–57.

Lu, Jane W. and Paul W. Beamish (2001) 'The Internationalization and Performance of SMEs', *Strategic Management Journal*, 22(6/7), 565–86.

Lumpkin, G. Tom and Gregory G. Dess (2001) 'Linking Two Dimensions of Entrepreneurial Orientation to Firm Performance: The Moderating Role of Environment and Industry Life Cycle', *Journal of Business Venturing*, 16(5), 429–51.

Madsen, Tage K., Erik S. Rasmussen and Per Servais (2000) 'Differences and Similarities between Born Globals and Other Types of Exporters', *Advances in International Marketing*, 10, 247–65.

Matthyssens, Paul and Pieter Pauwels (1996) 'Assessing Export Performance Measurement', *Advances in International Marketing*, 8, 85–114.

Moen, Oystein (2002) 'The Born Globals – A New Generation of Small European Exporters', *International Marketing Review*, 19(2), 156–75.

Moen, Oystein and Per Servais (2002) 'Born Global or Gradual Global? Examining the Export Behavior of Small and Medium-Sized Enterprises', *Journal of International Marketing*, 10(3), 49–72.

Narver, John C. and Stanley F. Slater (1990) 'The Effect of a Market Orientation on Business Profitability', *Journal of Marketing*, 54(5), 20–35.

Nunnally, Jum C. (1978) *Psychometric Theory* (New York: McGraw-Hill).

O'Grady, Shawna and Henry W. Lane (1996) 'The Psychic Distance Paradox', *Journal of International Business Studies*, 27(2), 309–33.

Ohmae, Kenichi (1989) 'Managing in a Borderless World', *Harvard Business Review*, 67(3), 152–61.

Oviatt, Ben M. and Patricia P. McDougall (1994) 'Toward a Theory of International New Ventures', *Journal of International Business Studies*, 25(1), 45–64.

Piercy, Nigel (1981) 'British Export Market Selection and Pricing', *Industrial Marketing Management*, 10(4), 287–97.

Porter, Michael (1991) 'Towards a Dynamic Theory of Strategy', *Strategic Management Journal*, 12 (Winter Special Issue), 95–117.

Rennie, Michael W. (1993) 'Global Competitiveness: Born Global', *The McKinsey Quarterly*, 4, 45–52.

Reuber, A. Rebecca and Eileen Fischer (1997) 'The Influence of the Management Team's International Experience on the Internationalization Behaviors of SMEs', *Journal of International Business Studies*, 28(4), 807–25.

Rialp, Alex, Josep Rialp and Gary A. Knight (2005) 'The Phenomenon of Early Internationalizing Firms: What Do We Know After a Decade (1993–2003) of Scientific Inquiry?', *International Business Review*, 14, 147–66.

Rugman, Alan M. and Alan Verbeke (2004) 'A Perspective on Regional and Global Strategies of Multinational Enterprises', *Journal of International Business Studies*, 35, 3–18.

Ruigrok, Winfried and Hardy Wagner (2003) 'Internationalization and Performance: An Organizational Learning Perspective', *Management International Review*, 43(1), 63–83.

SKOD (1998) *Scandinavian Kompass on Disc*, 1998(1). Published by Kompass Norway, Finland, Sweden and Denmark.

Snyder, William M. and Thomas G. Cummings (1998) 'Organization Learning Disorders: Conceptual Model and Intervention Hypotheses', 51, 873–95.

Sullivan, Daniel (1994) 'Measuring the Degree of Internationalization of a Firm', *Journal of International Business Studies*, 25(2), 325–42.

Tallman, Stephen and Jiatao Li (1996) 'Effects of International Diversity and Product Diversity on the Performance of Multinational Firms', *Academy of Management Journal*, 39(1), 179–96.

Tessler, Andrew (1980) 'Britain's Over-Ambitious Exporters', *Marketing U.K.*, February, 67–74.

Vermeulen, Freek and Harry Barkema (2002) 'Pace, Rhythm, and Scope: Process Dependence in Building a Profitable Multinational Corporation', *Strategic Management Journal*, 23(7), 637–53.

Yamin, Mohammad and Rudolf R. Sinkovics (2006) 'Online Internationalisation, Psychic Distance Reduction and the Virtuality Trap', *International Business Review*, 15(4), 339–60.

Zahra, Shaker A. and G. George (2002) 'International Entrepreneurship: The Current Status of the Field and Future Research Agenda', in Michael A. Hitt, R. Duane Ireland,

Donald L. Sexton and Michael S. Camp (eds), *Entrepreneurship: Creating an Integrated Mindset*, (Oxford: Blackwell), 255–88.

Zahra, Shaker A., Duane R. Ireland and Michael A. Hitt (2000) 'International Expansion by New Venture Firms: International Diversity, Mode of Market Entry, Technological Learning and Performance', *Academy of Management Journal*, 43(5), 925–50.

14

The Outcomes of Unsolicited International Enquiries Received by SMEs

Nan Sheng Zhang and Rod B. McNaughton

Introduction

Whether firms are predominately national or international in their market focus, they make investments of capital and management effort by advertising to strengthen brand recognition, reducing prices to attract more customers, or expanding their distribution network to increase penetration and market share. These strategies share a common characteristic: they 'solicit' customers in the target market and actively encourage them to buy. In contrast, unsolicited enquires come about without any active attempt to encourage the buyer. Unsolicited enquiries come from markets or segments that vendors are not actively targeting, and can reveal latent sources of unidentified demand. Liang (1995) defines unsolicited enquiries as a special case of organizational buying behaviour, where the buyer's search for a vendor is *ad hoc*, and the order is given to a foreign supplier who is not actively soliciting the buyer's business.

Unsolicited international enquiries are one of the key factors that stimulate the initiation of exporting (Aaby and Slater, 1989; Bilkey, 1978; Katsikeas and Piercy, 1993; Miesenbock, 1988). The internationalization literature portrays exporting in response to an unsolicited export order as the first stage of the internationalization process for firms (Barker and Kaynak, 1992), and identifies unsolicited enquiries as one of the most important external stimuli for the initiation of exports (Czinkota and Tesar, 1982). A number of empirical studies confirm the significance of unsolicited enquiries. For example, Bilkey (1978), da Rocha, *et al.* (1990) and Karafakioglu (1986) all found that more than 40 per cent of firms initiated their international activities based on unsolicited enquiries from overseas buyers.

The incidence of unsolicited enquiries increases when a firm establishes a presence on the World Wide Web. Compared to traditional media, the Internet has the ability to overcome barriers of distance, time and space. These characteristics allow customers to access company information easily from any location at any time of the day. The convenience of the Web encourages importers to search online and send their unsolicited enquiries directly. However, not all the literature is positive about unsolicited enquiries. Liang (1995), for example, suggests that unsolicited buyers are not necessarily reliable and valuable customers. Thus the issue of how to deal with unsolicited enquiries is a source of anxiety for managers of SMEs, especially when having a Web presence makes a firm easy to find by potential buyers located around the world.

How valuable are unsolicited enquiries from abroad? What proportion of enquiries are typically converted to completed sales, and do any of the buyers go on to become regular customers? The extant literature does not answer these questions adequately. Thus the objective of this chapter is to evaluate the quality of unsolicited international enquiries in terms of their outcomes. The study reports on the likelihood that an unsolicited order will result in a completed sale, and whether the sale generated continuing business for a sample of SMEs located in Scotland.

Literature review

Studies of export behaviour identify a number of motivations for initiating exports, such as loss of domestic market, home market saturation, size of foreign markets, home or host government incentives, and unsolicited foreign enquiries (Aaby and Slater, 1989; Bilkey, 1978; Katsikeas and Piercy, 1993; Miesenbock, 1988). Unsolicited enquiries from abroad are one of the key factors that stimulate the initiation of exporting. Cavusgil and Naor (1987) argue that many firms regard responding to unsolicited enquiries as a low-risk activity. From the exporter's point of view, unsolicited enquiries require little effort or investment, and the sales cycle is short. A number of empirical studies confirm the significance of unsolicited enquiries. Bilkey (1978) reviewed seven export studies and found that the proportion of firms that initiated their overseas business based on unsolicited enquiries ranged from 40 per cent to 83 per cent. da Rocha *et al.* (1990) found that 62 per cent of the firms in their sample started exporting in response to unsolicited enquiries, and Karafakioglu (1986) found a similar proportion (61 per cent) in his sample.

The stimuli firms receive during their pre-export stage is crucial, determining whether a firm begins to export or not, and the nature of its future approach to exporting (Welch and Wiedersheim-Paul, 1980). The frequency with which unsolicited enquiries spur interest in international markets led Bilkey and Tesar (1977) to conclude that receipt of an unsolicited enquiry is the critical factor affecting whether companies export experimentally or not. However, some scholars question the real effect of unsolicited enquiries. Leonidou (1995) reviewed thirty studies on export stimulation and concluded that various inconspicuous forces facilitate or inhibit the effective influence of stimuli such as unsolicited enquiries. Barker and Kaynak (1992) distinguished between factors responsible for exporting at the initiating exporting stage and the sustaining exporting stage. By comparing the importance of factors in the initiating and continuing stages, they demonstrated that unsolicited enquiries become less important as a firm gains experience. Barker and Kaynak (1992, p. 30) concluded that while receiving an unsolicited enquiry is 'very instrumental in increasing interest in exporting, once the firm is involved, the same factor is no longer as critical as in continuing to export'.

It is common to classify export stimuli as either internal or external. Internal stimuli are those associated with the firm's endogenous influences, while external stimuli are derived from the environment, such as where a firm locates (Kaynak and Stevenson, 1982; Ogram, 1982; Wiedersheim-Paul *et al.*, 1978). Lee and Brasch (1978) characterized an export decision stimulated by internal factors as rational, objective-orientated behaviour and with a problem-orientated adoption process. They described exporting motivated by external factors as being less rational, less objective-orientated, and with an innovation-orientated adoption process. Export stimuli can also be categorized into proactive factors and reactive factors. Proactive factors suggest that a firm begins to export because of its interest in taking advantage of a specific internal competitive advantage or market opportunity. Reactive factors imply that a firm begins to export in response to environmental pressures (Czinkota and Johnston, 1981; Czinkota and Tesar, 1982). Proactively motivated companies are characterized by an aggressive, positive and strategic approach towards exporting, while companies stimulated by reactive factors display passive, negative and tactical behaviour (Cavusgil, 1980; Czinkota and Johnston, 1981).

The internal/external and the proactive/reactive typologies are often tautological, with internal stimuli being perceived as proactive, and external stimulating factors considered reactive (Leonidou, 1995). However, this assumption is problematic – some internal stimuli show a reactive nature (for example, excess capacity or resources), while some external stimuli exhibit a purely proactive character (for example, the identification of foreign market opportunities). Thus, Albaum *et al.* (1989) suggested the typologies be combined into four categories: internal-proactive, external-proactive, internal-reactive and external-reactive. The exporters motivated by internal-proactive factors are likely to have active strategies, while those stimulated by external-reactive factors are inclined to be passive in their business operations. Unsolicited enquiries fall into the category 'external-reactive'.

Simply responding to unsolicited enquiries is a passive export strategy. Active exporters are motivated by more positive stimuli than are passive exporters (Czinkota and Tesar, 1982; Pavord and Bogart, 1975; Piercy, 1981). Moreover, many researchers (for example, Jaffe *et al.*, 1988; Koh, 1989) point out that active exporters are more likely to be successful in export performance than are reactive exporters. The assumption is that a well-prepared and highly stimulated firm will have a better chance of succeeding in international markets than an ill-prepared and weakly stimulated company (Welch and Wiedersheim-Paul, 1980). This argument implies that unsolicited enquiries are a poor base on which to establish export activities. However, there is little empirical evidence reported in the literature to test this hypothesis.

What are the characteristics of unsolicited enquiries, and of the buyers that initiate them? Most of the extant research in the international business literature focuses exclusively on the behaviour of exporters, not importers. Thus little is known about the characteristics of the importers who initiate unsolicited enquiries, or the outcomes of the enquiries. Liang (1995) identified this gap in the literature and researched unsolicited enquiries from the perspective of importers. He developed fourteen propositions of importer behaviour, which explain why and how importers choose the recipients of their unsolicited enquiries. The propositions range from the firm's search capability and the stages of their internationalization process to the company's network extension and the executives' personalities. Among these fourteen propositions, three are particularly relevant (Liang, 1995, p. 45):

- When systematic vendor search is beyond their capacity, importing organizations will adopt a sequential search approach and give unsolicited international orders to the first overseas supplier they find to be 'good enough'.
- Companies at the beginning stage of the internationalization process, with limited overseas connections, are more likely to place unsolicited international orders; whereas those with a long history of international presence and well-established overseas networks are more likely to take a more systematic vendor search approach.
- Companies are more likely to place unsolicited international orders in their initial entry into a sourcing country; later vendor search and selection within that country are more likely to follow the rational models.

The first proposition implies that some unsolicited enquiries come from SMEs, which often lack a vendor search capability. The second indicates that the buyers are usually just at the initiation stage of their internationalization path, and lack import experience. The third suggests that newcomers are more likely to initiate unsolicited enquiries. Liang (1995) does not test these propositions empirically, but if they are correct, the implication is that the quality of unsolicited enquiries may be generally poor, though not necessarily in all cases.

In sum, the literature identifies unsolicited enquiries as playing a significant role in stimulating exporting and initiating internationalization. However, some research questions the positive effect of such enquiries, hence the relative quality of unsolicited enquiries, and the outcomes of responding to such enquires, are an important issue for investigation.

Survey method

Our research used an e-mail survey to collect data. There is a growing body of research devoted to exploring and evaluating the practicability and quality of e-mail surveys (Ranchhod and Zhou, 2001). Some researchers suggest that e-mail has the advantage of enabling more interaction between researchers and respondents (Smith, 1997), and creating customized questionnaires for different respondents (Comley, 1996). Empirical studies report benefits of response quality, response speed and reduced survey cost (Ranchhod and Zhou, 2001). Therefore, Tse (1998) concluded that traditional, expensive, time-consuming marketing survey studies can be carried out via e-mail and be expected to achieve reasonable results. However, the new survey technology is not always perfect; it also poses several problems, to which particular attention should be paid (Taylor, 1999). One of the major problems is a low response rate compared with conventional research survey methods (Basi, 1999; Kent and Lee, 1999). The results of a number of studies indicate that the response rate from e-mail surveys is usually less than 10 per cent (Basi, 1999; Kent and Lee, 1999; Tse, 1995).

To conduct our research, a sample of over 2,000 Scottish firms was obtained from three directories: details of approximately 1,000 firms were obtained from the directory of *Scottish Enterprise*; over 600 firms were selected from the *Federation of Small Business Online Directory*; and a list more than 400 firms were contained in the *Dundee & Tayside Chamber of Commerce Directory*. The listings in these directories included the e-mail address of a contact person. Positions held by the contact person included owner, marketing manager and sales manager.

Not all industries are suitable for this research. For example, retail firms (grocery stores, bookstores and so on) and service firms (hotels, restaurant and so on) that only provide services to local customers are not appropriate to include in the sample. Thus most of the firms included in the sample are manufacturers, ranging from textiles and chemical producers to microelectronics and optoelectronics. After removing firms in inappropriate sectors, the number of firms in the target population was 1,257.

Each firm in the sample was sent an e-mail-based questionnaire regarding their experience with unsolicited enquiries. The method consisted of an initial e-mail, a confirmation and acknowledgement e-mail, follow-up mailing and an e-mail providing a summary of the results. The initial e-mail included a recruitment letter and a questionnaire. The recruitment letter was embedded in the initiating e-mail, while the questionnaire was included as an attachment. The e-mails were sent individually, as spam checkers typically use the number of recipients to trap messages.

Thirty-seven of the questionnaires were returned as undeliverable. Of the remaining 1,220 firms, 83 completed and returned the questionnaire, which resulted in a 6.8 per cent response rate. Follow-up mailings were made to all the non-respondents in order to determine the reasons for non-response and to increase the response rate. The most frequent reason given was that the firm managers did not have time to complete the questionnaire. Other reasons were that it is difficult for the firm to quantify such matters, or the respondents did not believe they could contribute much to the research because they received few, if any, unsolicited enquiries.

The questionnaire contained fourteen questions: five on the background characteristics of the firm (products, market, age, employees, and if they received unsolicited enquiries), and a further six about their experience with unsolicited enquiries. These later questions concerned the number of enquires received during the previous year, the proportion of enquiries that resulted in completed sales, and the proportion that became regular customers. Three additional questions gathered information on the proportion of enquiries generated from difference sources (including the Internet), and the outcome of Internet enquiries in particular.

Seventy-two valid questionnaires were obtained from the 83 responses. Of the 72 respondents, 68 had received unsolicited enquiries in the previous three years. Of the 68 firms, 18 were electronics firms, 27 were in the textiles field, 12 were chemical firms, 3 were machinery firms, and 8 were providers of other manufactured goods (for example, computers, gifts, and medical items). A Kruskal-Wallis one-way analysis of variance tested the hypothesis that there was no significant industry effect on the receipt of enquiries (the dependent variable was number of unsolicited enquiries). The test was not significant ($p < 0.01$). This test provided support for treating the sample as one group. More than half (56 per cent) of the firms indicated that their focus is the domestic (UK) market, and 12 per cent indicated that their focus is the local (regional UK) market. Thus, two-thirds of the firms in the sample are interested primarily in the domestic market, while a third are actively interested in international markets.

Table 14.1 presents the characteristics of the respondents in terms of firm age, number of employees and number of unsolicited enquiries received in the previous twelve months. The firms range in age from start-ups to those in their second or third generation of managers. There is a bias towards older 'survivors', as many startups are not yet captured in published directories. For the 68 firms in the sample, the number of unsolicited enquiries the firms received in the previous twelve months had the following characteristics: a range of 2 to 1,000, mean of 156, and standard deviation of 294. Some respondents did not

Table 14.1   Descriptive statistics of firm age and firm size

	N	Minimum	Maximum	Mean	St. Dev.
Age of firm (years)	68	1	115	19	26
Number of employees	68	1	300	41	64
Number of unsolicited international enquiries received in previous 12 months	68	2	1000	156	294

provide a specific number, but a range, such as from 80 to 100. In these cases, the mid-point of the range served as a point estimate. The data describe a sample of firms with considerable experience of receiving unsolicited international enquiries.

Results

The main objective of the research was to evaluate the quality of the unsolicited enquiries from international customers. That is, do unsolicited enquiries in fact result in sales, and how many become regular customers? Figure 14.1 shows the proportion of unsolicited enquiries received by respondents in the previous twelve months that resulted in completed sales. Most respondents (38.2 per cent) reported that none of the unsolicited enquiries they received resulted in a sale, and an additional quarter of respondents only completed a sale for one in ten unsolicited enquiries. The weighted mean of the distribution is 16.6 per cent, suggesting that, on average, only seventeen of every 100 unsolicited enquiries resulted in a completed sale.

Unsolicited enquiries that result in a sale may lead to further sales, and unsolicited buyers may become regular customers. Table 14.2 shows that almost 70 per cent of the respondents stated that none of their unsolicited buyers (that is, enquiries that converted to a sale) became regular customers. However, a sizeable minority (31.9 per cent) did establish a long-term relationship with some of their unsolicited buyers. The weighted mean is 4.5 per cent, indicating that, on average, only about five out of every 100 unsolicited enquiries that result in a sale are converted into regular customers who make additional purchases.

Respondents were asked about the source of unsolicited enquiries. Enquiries almost universally came through a website or e-mail (on average 96.6 per cent, with a range between 70–100 per cent). Respondents were asked to estimate the proportion of enquiries that came through the Internet for each category: a waste of time; no sale but learned something; a one-off sale; a sale that might lead to future sales; and the buyer became a regular customer. Table 14.3

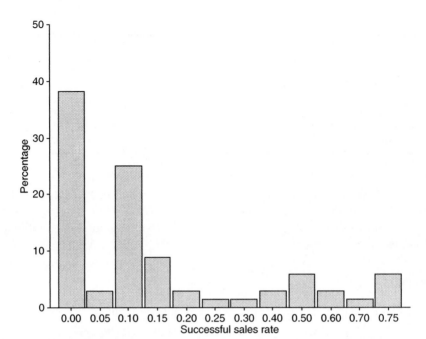

Figure 14.1   Distribution of unsolicited enquires converted to sales

presents the results. The majority of respondents reported having unsolicited enquiries that are a 'waste of time', and estimated that about 81 per cent of enquiries fall into this category. In contrast, only seven respondents had enquires that resulted in an ongoing relationship (and an additional seven

Table 14.2   Conversion rates from one-off sale (resulting from unsolicited enquires) to regular customer

Rate of conversion to regular customer (A)	Number of respondents	Percentage (B)	A × B (%)
0.00	47	69.1	0
0.05	7	10.3	0.5
0.10	5	7.4	0.7
0.15	2	2.9	0.4
0.20	1	1.5	0.3
0.25	2	2.9	0.7
0.30	4	5.9	1.8
Total	**68**	**100.0**	**4.4**

Table 14.3 The results of unsolicited enquiries received through the World Wide Web

Result	Number of respondents having enquiries in this category	Mean percentage of enquiries falling in category
A waste of our time	55	80.9
We did not make a sale, but we learned something	48	70.6
We made a one-off sale	28	41.2
The buyer may purchase from us again in the future	7	10.3
The buyer is likely to become a regular customer	7	10.3

reported sales that might lead to further sales), with about 10 per cent of their enquiries falling into this category.

Conclusion

The literature on international marketing emphasizes the importance of various stimuli to the export behaviour of firms (for example, Leonidou, 1995). Unsolicited enquiries from international customers are characterized as being less risky and a major stimulus for exports (for example, Aaby and Slater, 1989; Bilkey, 1978; Katsikeas and Piercy, 1993; Miesenbock, 1988). The wide adoption of the Internet has greatly reduced the cost of finding, learning about and contacting potential vendors overseas, thus increasing the likelihood and number of unsolicited enquiries that firms receive. The literature also contains contrasting opinions about the longer-term advantages of unsolicited enquiries, and about the quality of such customers. For example, Barker and Kaynak (1992) find unsolicited foreign enquiries are not critical at the sustained exporting stage, and others (for example, Liang, 1995) suggest that unsolicited enquiries are not necessarily reliable and valuable.

The objective of this chapter is to evaluate the quality of unsolicited overseas enquiries. The most practical and direct means of assessing the quality of an unsolicited enquiry is to analyse the result: whether the enquiries result in completed transactions, and if they lead to future sales. Data are analysed that report the experiences of 68 Scottish SMEs, reporting on average 156 unsolicited enquires over the previous twelve months. Over a third of the firms (38.2 per cent) reported that none of the unsolicited enquiries they received resulted in a completed sale, and a further quarter of firms reported that only a tenth of enquires are converted to sales. Nearly 70 per cent of respondents claimed that none of those who were converted to buyers followed up with further orders.

Finally, more than 90 per cent of enquiries are generated from an online presence, and the respondents classified the vast majority of these as 'a waste of time'.

While unsolicited enquiries are credited with reducing selling time and costs, and possibly resulting in more satisfied and loyal customers, these results suggest the opposite: the respondents attracted a lot of attention, especially through their Web presence, but a relatively small proportion of these contacts turn into sales, and an even smaller proportion into an ongoing relationship.

Implications for SMEs

The first implication concerns the expectation of unsolicited enquiries. Unsolicited enquiries are passive, so little investment goes into generating them. Further, those who enquire have self-identified as potential customers. Thus unsolicited enquiries are a potential source of cost-effective leads. There is a bias towards seeing unsolicited enquires as positive for sales development, especially in attracting a firm to enter foreign market that it might otherwise have overlooked. However, our research suggests that unsolicited orders rarely result in completed sales and ongoing customer relationships. Responding to such enquiries (especially given the number that some firms reported) probably takes a considerable amount of time, and firms should evaluate the benefits versus costs of the resources committed to this. In some cases, it may be possible to automate responses, or at least to screen enquiries by collecting more information before the query is forwarded for an individual response. A second implication concerns actions to improve the success rate of converting unsolicited enquiries into sales. While not identified directly by this research, it is logical that potential influences include website design, the level of detail and self-service mechanisms provided, the way in which enquiries are collected (and how much information is requested), search engine placement, affiliate and referral links, and the method of responding to enquires. There is also an implication for potential importers – while the Internet reduces search costs, it also increases the number of vendors to be considered, and makes vendors in other countries appear more accessible. Refined searches and a systematic process to evaluate potential vendors from Web-based information reduces costs for both buyer and potential vendor.

Limitations and future research directions

Limitations of the current study include the relatively small sample size and low response rate, and the selection of firms from a small economy. Unsolicited enquiries are a topic of considerable importance in the international business literature, in particular with regard to export initiation. While this was a small-scale study, it is important because it provides empirical evidence of the volume and outcome of such enquiries, as well as the distribution of their outcomes. Studies that provide additional empirical evidence would be welcome additions to this stream of literature. In addition, insight into what attracts enquiries, how the enquiries are handled, and how to improve the attraction of qualified buyers and conversion rates is valuable, and notably absent from the extant literature. Related research could compare enquiries from local and more distant domestic firms with those from international buyers, to compare conversion rates, identify differences, and estimate the extent to which location influences the receipt and outcome of unsolicited enquiries.

References

Aaby, N.-E. and S. F. Slater (1989) 'Management Influences on Export Performance: A Review of the Empirical Literature 1978–88', *International Marketing Review*, 6(4), 7–26.

Albaum, G., J. Strandskov, E. Duerr, and L. Dowd (1989) *International Marketing and Export Management*, (Reading, Mass.: Addison-Wesley).

Armstrong R. W. and J. E. Everett (1993) 'Managerial Perceptions of Ethical Problems in International Marketing: Australian Evidence', *Asian Journal of Marketing*, December, 61–71.

Barker, A. T. and E. Kaynak (1992) 'An Empirical Investigation of the Differences Between Initiating Exporters', *European Journal of Marketing*, 26(3), 27–36.

Barrett, N. J. and I. F. Wilkinson, (1986) 'Internationalisation Behaviour: Management Characteristics of Australian Manufacturing Firms by Level of International Development', in P. W. Turndull and S. J. Paliwoda (eds), *Research in International Marketing* (London: Croom Helm) 213–33.

Basi, R. K. (1999) 'World Wide Web Response Rates to Socio-demographic Items', *Journal of the Market Research Society*, 41(4), 397–401.

Bennett, R. J. and C. Smith (2002) 'Competitive Conditions, Competitive Advantage and the Location of SMEs', *Journal of Small Business and Enterprise Development*, 9(1), 73–86.

Bilkey, W. J. (1978) 'An Attempted Integration of the Literature on the Export Behavior of Firms', *Journal of International Business Studies*, 9(1), 33–46.

Bilkey, W. J. and G. Tesar (1977) 'The Export Behaviour of Smaller Wisconsin Manufacturing Firms', *Journal of International Business Studies*, 8, 93–98.

Blattberg, R. C. and R. Glazer (1993) *The Marketing in Information Revolution*. (Boston, Mass.: Harvard Business School Press).

Cateora, P. R. and P. N. Ghauri (2000) *International Marketing* (London: McGraw-Hill).

Cavusgil, S. T. (1980) 'On the Internationalization Process of Firms', *European Research*, 8(6), November, 273–81.

Cavusgil, S. T. and J. Naor (1987) 'Firm and Management Characteristics as Discriminators of Export Marketing Activity', *Journal of Business Research*, 15, 221–35.

Cohen, D., J. G. March and J. P. Olsen (1972) 'A Garbage Can Model of Organizational Choice', *Administrative Science Quarterly*, 17, 1–25.

Comley, P. (1996) 'The Use of the Internet as a Data Collection Method', http:// www.virtualsurveys.com/news/papers/paper_9.asp (accessed 23 August 2004).

Czinkota, M. R. and W. J. Johnston (1981) 'Segmenting US Firms for Export Development', *Journal of Business Research*, 9, 353–65.

Czinkota, M. R. and G. Tesar (1982) *Export Management: An International Context* (New York: Praeger).

da Rocha, A., C. H. Christensen and C. E. da Cunha (1990) 'Aggressive and Passive Exporters: A Study of the Brazilian Furniture Industry', *International Marketing Review*, 7(5), 6–15.

Dichtl, E., H. -G. Koglmayr and S. Muller (1990) 'International Orientation as a Precondition for Export Success', *Journal of International Business Studies*, 21(1), 23–40.

Dichtl, E., M. Leibold, H. -G. Koglmayr and S. Muller (1984) 'The Export Decisions of Small and Medium-Sized Firms: A Review', *Management International Review*, 24(2), 49–60.

Ellsworth, J. H. and M. V. Ellsworth (1996) *Marketing on the Internet – Multimedia Strategies for the WWW* (New York: John Wiley).

Emery, V. (1995) *How to Grow your Business on the Internet* (Scotsdale, Ariz.: Coriolis Books).

Hamill, J (1997) 'The Internet and International Marketing', *International Marketing Review*, 14(5), 300–23.

Jaffe, E., D. H. Pasternak and I. Nebenzahl (1988) 'The Export Behaviour of Small Israeli Manufacturers', *Journal of Global Marketing*, 2(2), 27–49.

Johanson, J. K. (2003) *Global Marketing: Foreign Entry, Local Marketing, and Global Management* 3rd edn (New York: McGraw-Hill).

Johanson, J. K. and F. Wiedersheim-Paul (1975) 'The Internationalization of the Firm: Four Swedish Cases', *Journal of Management Studies*, 12: 305–22.

Kalakata, R. and M. Robinson (2001) *E-business 2.0 Roadmap for Success* (Boston, Mass.: Addison-Wesley).

Karafakioglu, M. (1986) 'Export Activities of Turkish Manufacturers', *International Marketing Review*, 3(4), 34–43.

Katsikeas, C. S. and N. F. Piercy (1993) 'Long-term Export Stimuli and Firm Characteristics in a European LDC', *Journal of International Marketing*, 1(3), 23–47.

Kaynak, E. and L. Stevenson (1982) 'Export Orientation of Nova Scotia Manufacturers', *Export Management: An International Context* (New York: Praeger), 132–45.

Kent, R. and M. Lee (1999) 'Using the Internet for Market Research: A Study of Private Trading on the Internet', *Journal of the Market Research Society*, 41(4), 377–85.

Koh, A. C. (1989) 'An Evaluation of the Current Marketing Practices of United States Firms', *Developments in Marketing Science*, 7, 198–203.

Kotabe, M. and M. Czinkota (1992) 'State Government Promotion of Manufacturing Exports: A Gap Analysis', *Journal of International Business Studies*, 23, 637–58.

Lee, W. and J. J. Brasch (1978) 'The Adoption of Exports as an Innovative Strategy', *Journal of International Business Studies*, 9(1), 85–93.

Leonidou, L. C. (1995) 'Export Stimulation Research: Review, Evaluation and Integration', *International Business Review*, 4(2), 133–56.

Liang, N. (1995) 'Soliciting Unsolicited Export Orders: Are Recipients Chosen at Random?', *European Journal of Marketing*, 29(8), 37.

Mettrop, W. and P. Nieuwenhuysen (2001) 'Internet Search Engines – Fluctuations in Document Accessibility', *Journal of Documentation*, 57(5), 623–51.

Miesenbock, K. J. (1988) 'Small Businesses and Exporting: A Literature Review'. *International Small Business Journal*, 6(2), 42–61.

Moon, J. and H. Lee (1990) 'On the Internal Correlates of Export Stage Development: An Empirical Investigation in the Korean Electronics Industry', *International Marketing Review*, 7(5), 16–26.

Morgan, R. E. and C. S. Katsikeas (1997) 'Export Stimuli: Export Intention Compared with Export Activity', *International Business Review*, 6(5), 477–99.

Ohmae, K. (1990) *The Borderless World* (London: William Collins).

Olson, H. C. and F. Wiedersheim-Paul (1978) 'Factors Affecting the Pre-Export Behaviour of Non-Exporting Firms', *European Research in International Business* (New York: North-Holland), 283–305.

Olson, P. D., N. Gough and D. W. Bokor (1997) 'Export Planning and Performance: An Organizational Culture Perspective on Small Firms', *http://www.usasbe.org/knowledge/ proceedings/1997/P177Olson.PDF* (accessed 23 August 2004).

Orgam, E. W. (1982) 'Exporters and Non-Exporters: A Profile of Small Manufacturing Firms in Georgia', in M. R. Czinkota and G. Tesar (eds), *Export Management: An International Context* (New York: Praeger), 70–84.

Pavord, W. C. and R. G. Bogart (1975) 'The Dynamics of the Decision to Export', *Akron Business and Economic Review*, 6, 6–11.

Piercy, N. (1981) 'Company Internationalization: Active and Reactive Exporting', *European Journal of Marketing*, 15(3), 26–40.

Ranchhod, A. and Zhou, F. (2001) 'Comparing Respondents of E-mail and Mail Surveys: Understanding the Implications of Technology', *Marketing Intelligence & Planning*, 19(4), 254–62.

Robinson, P. J., C. W. Faris and Y. Wind (1967) *Industrial Buying and Creative Marketing* (Boston, Mass.: Allyn & Bacon).

Samiee, S. (1998) 'Exporting and the Internet: A Conceptual Perspective', *International Marketing Review*, 15(5), 413–26.

Schuldt, B. A. and J. W. Totten (1994) 'Electronic Mail versus Mail Survey Response Rates', *Marketing Research*, 6(1), 36–9.

Shenkar, O. and Y. Luo (2003) *International Business* (New York: Wiley).

Smith, C. B. (1997) 'Casting the Net: Survey of an Internet Population', *Journal of Communication Mediated by Computers*, 3(1), 42–63.

Taylor, H. (1999) 'Does Internet Research Work? Comparing On-line Survey Results with Telephone Surveys', *International Journal of Market Research*, 42(1), 51–63.

Tse, A. C. B. (1995) 'Comparing Two Methods of Sending Out Questionnaires: E-mail versus Mail', *Journal of the Market Research Society*, 37(4), 441–6.

Tse, A. C. B. (1998) 'Comparing the Response Rate, Response Speed, and Response Quality of Two Methods of Sending Questionnaires: E-mail vs. Mail', *Journal of the Market Research Society*, 40(4), 353–61.

Webster, F. E. and Y. Wind (1972) 'A General Model for Understanding Organizational Buying Behavior', *Journal of Marketing*, 36, 12–19.

Welch, L. S. and F. Wiedersheim-Paul (1980) 'Initial Exports – A Marketing Failure?', *Journal of Management Studies*, 17, 333–44.

Westhead, P., M. Binks, D. Ucbasaran and M. Wright (2002) 'Internationalization of SMEs: A Research Note', *Journal of Small Business and Enterprise Development*, 9(1), 38–48.

Wiedersheim-Paul, F., H. C. Olson and L. S. Welch (1978) 'Pre-export Activity: The First Step in Internationalization', *Journal of International Business Studies*, 9(1), 47–58.

Wind, Y. (1970) 'Industrial Source Loyalty', *Journal of Marketing Research*, 9, 450–7.

Young, S. (1995) 'Export Marketing: Conceptual and Empirical Developments', *European Journal of Marketing*, 29(8), 7–16.

Index

Notes: f = figure; n = note; t = table; **bold** = extended discussion or heading emphasized in main text.

257